Antimatter

Antimatter

FRANK CLOSE

OXFORD

UNIVERSITY PRESS

Great Clarendon Street, Oxford OX2 6DP

Oxford University Press is a department of the University of Oxford.
It furthers the University's objective of excellence in research, scholarship,
and education by publishing worldwide in

Oxford New York

Auckland Cape Town Dar es Salaam Hong Kong Karachi
Kuala Lumpur Madrid Melbourne Mexico City Nairobi
New Delhi Shanghai Taipei Toronto

With offices in

Argentina Austria Brazil Chile Czech Republic France Greece
Guatemala Hungary Italy Japan Poland Portugal Singapore
South Korea Switzerland Thailand Turkey Ukraine Vietnam

Oxford is a registered trademark of Oxford University Press
in the UK and in certain other countries

Published in the United States
by Oxford University Press Inc., New York

© Frank Close 2009

The moral rights of the author have been asserted
Database right Oxford University Press (maker)

First published 2009

British Library Cataloguing in Publication Data

Data available

Library of Congress Cataloging in Publication Data

Typeset by SPI Publisher Services, Pondicherry, India
Printed in Great Britain
on acid-free paper by
Clays Ltd, St Ives plc

ISBN 978-0-19-955016-6 (Hbk.)
       978-0-19-957887-0 (Pbk.)

5 7 9 10 8 6

Acknowledgements

In thirty years of lecturing and broadcasting about science, I have been asked about antimatter more often than any other topic. On 4 October 2007 I appeared on *In Our Time* on BBC Radio 4 discussing antimatter with Melvyn Bragg, Val Gibson, and Ruth Gregory. That broadcast led to many emails and letters requesting news about antimatter. Among these was a disturbing new feature: the belief that antimatter would make weapons, create awful destruction, and that work by the US military had inspired Dan Brown's book *Angels and Demons*, which stars an antimatter bomb made, allegedly, at CERN.

These events inspired me to write this in the hope of setting the record straight, and of answering the questions that keep recurring. In the process I learned a lot about the reality of the military work, and the fanciful nonsense that created some of the hype. Although I have penned the words, they are the result of conversations with many colleagues over many years. In particular I am indebted to Rolf Landua for checking my numbers, correcting several mistakes in my early drafts, and lengthy discussions about antimatter. Betsy Devine, George Kalmus, Michael Marten, and my editor Latha Menon read my drafts in part or in whole, and their suggestions were incorporated more often than not. I am indebted to Gerald Smith for copies of his papers on antimatter research, to Stan Brodsky and Thornton Greenland for discussions on stretched positronium, to Kathryn Maris for showing me poetry and who compared matter and antimatter thus: 'Cain and Abel are siblings; their parents are the Original Parents (Big Bang); and one sibling kills off the other', and to the many colleagues at CERN and Oxford whose unrecorded comments have influenced what I have written.

Contents

Foreword

Genesis

In the beginning, there was nothing; 'there was darkness on the face of the void'. Then came a burst of energy: 'let there be light and there was light', though from where it came I don't know.

What we do know is what happened next: this energy coagulated into matter and its mysterious opposite, antimatter, in perfect counterbalance. Ordinary matter is the familiar stuff, which makes air, rocks, and living things. But matter's faithful opposite, identical in all respects except that deep inside its atoms everything is back to front, is unfamiliar. This is antimatter—the antithesis to matter.

Today antimatter does not exist normally, at least on earth, a vanishing act that is one of the unexplained mysteries of the universe. But we know that antimatter is real because scientists have made small pieces of it.

Antimatter destroys any matter that it touches in a pyrotechnic flash, an explosive release of all the energy that had been locked within for billions of years. Antimatter thus could become a wonderful source of power, the technology of the 21st century. Or instead, its potential to consign matter to oblivion could make antimatter the ultimate weapon of mass destruction.

At least, that is what popular literature, bloggers, and even some in the US Air Force believe. Could this really be true?

1

ANTIMATTER: FACT OR FICTION?

'What happens when the irresistible force meets the immovable object?' My father didn't beat about the bush when it came to the mysteries of the universe, and as Isaac Newton hadn't been satisfied with just one law of motion, or Beethoven with one symphony, so Dad had more than one question: 'How would you store a substance that destroyed everything it touched?'

The idea of something that would first consume its container had awesome implications: why stop there? Having destroyed its prison and escaped, it would then be free to devour the surroundings, condemning everything in its way into oblivion, including eventually each of us. This would be a truly irresistible force, the stuff of nightmares and horror fiction.

I decided that there was the answer: his questions were about a fiction. I was wrong.

Irresistible forces meeting immovable objects—that is a fiction that plays on the concept of the infinite: while philosophers might contemplate the paradoxes raised by two competing infinities, they are resolved by 'my infinity is bigger than yours'. However, the all-consuming substance is another matter, literally so, known as antimatter. Beloved by writers of fiction, its existence is nonetheless real, its implications profound.

Antimatter is a weird topsy-turvy shadow of matter, like tweedledum to our tweedledee, where left becomes right and positive turns into negative. Like the mould that remains when the cast is removed, matter and antimatter are the yin and yang of reality. When a child on the seashore digs a hole in the wet hard flat sand to build a sandcastle, the castle is a metaphor for matter and the hole for antimatter.

Were any material substance to meet its antimatter doppelganger, their complementary characteristics would mutually cancel one another in a dance of death. Antimatter's pyrotechnic ability to destroy matter in a flash of light is the source of its fascination. Antimatter is truly anti-matter.

It is eighty years since the possible existence of such a weird stuff was first proposed and three-quarters of a century since the first example, known as a 'positron', was seen.

We are still here, happily, because antimatter is extremely rare, almost nonexistent in fact, and when a piece sends matter to oblivion, it too is destroyed. That first seen speck of antimatter is long gone, having annihilated but a single electron in one atom. In the entire universe, as far as we can tell, matter and not antimatter is the norm. It seems that the destruction of antimatter was one of the first acts after the Big Bang. The material universe that survives today contains the left-over remnants of a long-past Great Annihilation between antimatter and matter, the relic of which is electromagnetic radiation, the 'microwave background' that fills the cosmos fourteen billion years after that stupendous event. The bad witch is dead; matter has won; in the counterbalanced infinities of matter and antimatter, it was matter whose infinity was the larger.

But if some escaped, is still lurking out there somewhere in the vastness of the universe, and we came across it in our wanderings through space; or if some of it rained down from the heavens in cosmic rays: what then? A volume no bigger than the trunk of a car, were it filled with antimatter, could make a blast visible around the world. If that were to happen, then antimatter would be a real example of my father's question, but thankfully it does not have the awful implications of spreading and destroying like some all conquering army that my imagination had conjured with. Antimatter will destroy matter indeed, but at the expense of destroying itself; it is like a cancer that in killing its host has self-destructed. For each piece of matter in our world that is annihilated, the price for antimatter is that one piece of it too disappears. The result may be an explosive flash of radiation, gamma rays, which fly away from the battle site at the speed of light, but the threat from the antimatter has passed. That is why no antimatter survives, at least around here: all has been destroyed by the victorious matter.

So antimatter is not like some awful version of 'ice-nine', the fictional version of water in Kurt Vonnegut's novel *Cat's Cradle* that turned all liquid that it met into frozen solid. First a few puddles of water, then the streams, rivers, and finally the oceans of the world froze 'in a great vvraoomph'. Antimatter's range is limited. Nonetheless whatever it touches and destroys releases energy more explosively than anything else that we know.

Did Antimatter Hit the Earth?

If antimatter exists elsewhere in the universe, then once in a while you would expect some to hit the earth. If this has happened in our four billion year history, all signs will have long gone; meteorites leave craters around which the extraterrestrial material can still be found, but antimatter would have been destroyed in a flash. The only evidence of an antimatter strike would have been the devastating explosion when it struck and until the last millions of years no one would have been around to tell the tale. However, just a hundred years ago, in June 1908, something happened that has never been completely explained and which aficionados insist was the most recent example of a collision with extraterrestrial antimatter.

A thousand miles east of Moscow, stretching from the Arctic Sea in the north to Mongolia in the south, and from the Urals to Manchuria, is a sparsely inhabited region larger than the whole of western Europe. In the remote heart of this lonely continent is the hidden valley of the Tunguska river, named after the Tungus people, a small ethnic group that survived by hunting bears and deer in the forests where in the summer reindeer graze among the endless pine trees.

The day of 30 June 1908 dawned cloudless and sunny. At eight o'clock in the morning Sergei Semenov, a farmer, was sitting on the steps of his house when there was a huge explosion in the sky. He later told scientists that the fireball was so bright that it even made the light of the sun appear dark and so hot that his shirt 'was almost burnt on my body' and melted his neighbour's silverware.[1] Even more remarkable was that, when scientists later investigated, they realized that the explosion had happened nearly 60 kilometres away from Semenov. Another farmer, Vassili Ilich, said that there was a huge fire that 'destroyed the forest, the reindeer and all other animals'. When he and several neighbours went to investigate, they found the charred remains of some of the deer but the rest had completely disappeared.

The dazzling fireball crossed from the south-east to north-west in a matter of seconds. Seismic waves were detected around the globe, and pressure waves in the atmosphere spread throughout Russia and Europe. The blast was visible 700 kilometres away, and threw so much smoke and dust into the stratosphere that sunlight was scattered from the bright side of the globe right around into the earth's shadow. A quarter of the way round the world in London, daylight came early as the midnight sky was as light as early evening. Had the event happened in the USA over Chicago, the light flash would have been visible as far away as Tennessee, Pennsylavania, and Toronto, while the thunder would have been heard on the East Coast, as far south as Atlanta and west to the Rocky Mountains. Two months passed before normality had returned.

Something from outer space had hit the atmosphere. Such things have happened in the past, as shown for example by the huge meteor crater in Arizona, which was the result of a lump of rock, a small asteroid, hitting the earth. However the Tunguska event, as it has become known, was different, as became apparent years later when the first adventurous investigators, led by the Czechoslovakian scientist Leonid Kulik in 1927, reached the remote site. Had it been an asteroid, a lump of rock from the solar system that had smashed into the earth, then some tell-tale hole in the ground should have been there. However, there was no sign of any crater. They discovered that immediately below the explosion was a vast mud plain as if a thousand bulldozers had cleared the forest to prepare the foundations for a city the size of London.

Surrounding this bleak scene was a ring of charred tree stumps. Beyond this the trees lay scattered like matchsticks, felled by a tumultuous hurricane, the blastwave from the explosion. Life had been totally destroyed, and remained so for over a quarter of a century. Since then, the ground has been excavated to depths of over 30 metres, but no signs of meteorite material or any physical trace of the invader have ever been found.

Whatever hit the Earth that day had vanished into thin air. In 1965 a trio of scientists comprising a physicist, a chemist, and a geophysicist examined all the evidence in the hope of determining once and for all what had happened. Examination of occasional trees that had remained standing showed traces of the blast wave that had hit. This gave an idea of the strength of the winds; the energy required to make trees burn can also be computed. There were records that showed the earth's magnetic field was disturbed, and seismometers had recorded the strength of the apparent earthquake.

Reports of the brightness of the flash and its duration were then factored in to the calculation. They deduced that almost a million billion joules of energy had been released in a few seconds, which is similar to an hour's energy consumption by the entire United Kingdom,[2] and consistent with a nuclear explosion.

While a man-made nuclear explosion would have been a natural suspect today, it would not have been in 1908 when nuclear physics, as we know it, was still decades in the future. If the nuclear seeds of matter were indeed involved in the catastrophe, some natural cause would be called for. Prima facie pieces of evidence in the form of the blast and the singular lack of any material remnants at the scene were all consistent with antimatter in the form of a lump of antirock as small as a metre across having been responsible, destroying everything including the nuclei of atoms. I shall examine the forensic evidence later, after we have learned what is known about antimatter.

Powerful Antimatter

The size of the Tunguska event is a reminder of antimatter's latent power. If a lump of matter is your fuel, then antimatter is the spark that will release its energy in ways that, theoretically at least, cannot be bettered in nature.

The formation of matter in the Big Bang involved vast amounts of energy congealing into the particles that make up the atoms from which everything on earth is made. Chemical and nuclear reactions involve the rearrangement of those pieces in ways that free some of that inner energy, but even in the most violent explosions only trifling amounts are actually released compared to what was locked into matter at its birth billions of years ago.

Living things are chemical factories, liberating energy from the reactions between carbon, oxygen, and other elements from which they are made. The difference between your bodily warmth and the power of an explosive blast wave is primarily one of the timescales involved. In our bodies energy is released gradually as warmth, maintaining a temperature of about 37 degrees Centigrade in a healthy individual, or slightly hotter when reactions run faster to combat unwanted invaders such as a virus in a fever. A chemical explosion is no different in essence, it is just the same but faster. A decent meal will keep you going for hours whereas if the timescales are compressed such that the energy is given out in a millisecond, the results would be literally explosive.

Dramatic though they can be, conventional rockets and even the most powerful chemical explosions liberate only a billionth of the energy that is locked within atoms. Most of an atom's energy is stored within its nucleus, and when the nuclear spark is lit we have the power of Hiroshima and Nagasaki, which dwarfs that of chemical explosives. Yet even here only one part in a thousand of what is available is being released. Even fusion reactions—such as power the sun or the hydrogen bomb, which are the most powerful explosions known—use up only about one per cent of the total energy within matter. To release the lot we need to reverse the processes that congealed that energy into matter long ago.

That is what antimatter can do. Annihilating a kilogram of antimatter will give out about ten billion times the amount of energy released when a kilogram of TNT explodes. Per kilogram of fuel, this is also 1,000 times more energy than nuclear fission and 100 times more than nuclear fusion could generate.

Herein lies the fascination of antimatter for science fiction, where it is the ultra-efficient power source for spacecraft as in Star Trek. It has been a theme of real blue-sky thinking in energy research programmes at NASA. It would also raise the spectre of the ultimate weapon, the antimatter bomb. If Hiroshima and Bikini Atoll have shown us what a mere one part in a thousand of matter can do, courtesy of $E = mc^2$, then the consequences of liberating the lot would be unimaginable.

Perhaps we should not be surprised then by an article in the *San Francisco Chronicle* in October 2004, which broke the news that 'The U.S. Air Force is quietly spending millions of dollars investigating ways to use a radical power source—antimatter, the eerie "mirror" of ordinary matter—in future weapons'. The story spread around the world and in India escalated to the claim that not just the US Air Force but 'defence scientists in many countries are working on anti-matter weapon systems' that are 'small enough to hold in one's hand'.

Antimatter is very popular in science fiction, but it is also very real and here it seems that the military are developing antimatter weapons in fact. One of my primary goals in this book will be to attempt to separate fact from fiction in the antimatter story.

Antimatter Secrets

If the reports on the latest military adventurism are correct, the US Air Force is developing antimatter weapons. The stories seem to have grown out of a speech given on 24 March 2004 by Kenneth Edwards, director of the 'revolutionary munitions' team at the Munitions Directorate at Eglin Air Force Base in Florida. He was a keynote speaker at the NASA Institute for Advanced Concepts (NIAC) conference in Arlington, Virginia, and in that talk, Edwards discussed the potential uses of positrons—basic particles of antimatter. There is no doubt that Edwards was fully aware of and impressed by the potential of antimatter. His speech, which 'almost defied belief' according to some media reports, stressed that even specks of antimatter too small to see could be devastating. As an example 50-millionths of a gram of positrons would be enough to generate a blast equal to the explosion (roughly 4,000 pounds of TNT, according to the FBI) at the Alfred P. Murrah Federal Building in Oklahoma City in 1995, which killed 168 people and injured over 500.

Readers of the newspaper reports were reminded that these weapon systems 'are devastating' and that 'the level of destruction is unimaginable'; there is no 'will be' or 'could', only 'are' and 'is', as if these devices are already being developed. Antimatter weapons were presented as environmentally friendly: in contrast to regular nuclear bombs, positron bombs 'won't eject plumes of radioactive debris',[3] and the primary product of the annihilation of positrons and electrons was advertised as an invisible but extremely dangerous burst of gamma radiation, which 'can kill a large number of soldiers without touching the civilian population'.

When journalists from the *San Francisco Chronicle* started asking questions, the Air Force allegedly 'forbade its employees from publicly discussing the antimatter research program'. For conspiracy theorists this is the proof that the stories are true; that antimatter weapons of devastating power are in hand (metaphorically at least!).

What is the reality behind these claims? Are they feasible in principle let alone in practice? Is there any more to this than claims that Saddam Hussein was developing cold fusion weapons at the time of the first Gulf War?*

The US Air Force and other arms of the US government do have a reputation for researching bizarre ideas in the hope that 'if it is possible, then let it be us that do it'. As a practising high energy physicist, it would be disingenuous of me not to admit that, following the development of radar and the atomic bomb, government support for radio astronomy, nuclear and particle physics in the 1950s was not entirely driven by pure motives. Having seen the power that science had managed to unleash from the atomic nucleus, and with fusion (hydrogen) bombs already being developed, the cold war was a time for funding blue-sky ideas in science and technology lest the Soviet Union be first to develop the 'next big thing'. Alongside the more sober ideas were

---

* This claim was made at the time my book exposing the fraud of Cold Fusion appeared, by unfortunate chance being published the same day that the war began. The BBC pulled interviews fearing them to be too sensitive. The *New York Times* was more robust and ran the exposé on its front page.

others that verged on charlatanism. Telepathy, psychokinesis, and antigravity paint were but three, so it is plausible that governments have considered the possibilities for antimatter energy sources or antimatter weapons. Unlike the three examples just mentioned, there is good evidence for antimatter, as good as that for nuclear fission in 1939; the subsequent development of the atomic bomb was a tour de force of applied science and engineering. The successful construction of nuclear weapons confirmed the 'can do' approach for strategists in the USA. So antimatter devices at first sight appear to fit the bill.

It has been claimed that the US Air Force has been funding numerous scientific studies of the basic physics of antimatter for up to fifty years. More likely is that the advances made in antimatter research at open laboratories such as CERN in Europe and Fermilab in the USA, which began to hit the headlines after 1996, set the military in motion. In chapters 2 to 8 we will learn about the nature of antimatter, its history, the opportunities it presents, and also its limitations. With these insights, we will in the final chapter return and evaluate the claims about antimatter weapon projects.

Natural Antimatter

Although antimatter in bulk—even extremely small bulk—doesn't exist hereabouts, nonetheless some natural processes produce fleetingly the simplest example, the positron, the antiworld's mirror of the electron. As the electron, the lightest electrically charged particle, is found in the atoms of all matter, so the positron, its antimatter counterpart, is potentially an essential piece of anti-atoms in the antiworld. In our world many elements are radioactive, the nuclei of their atoms spitting out energy spontaneously as their constituent pieces rearrange themselves to form more stable assemblies. The atomic nuclei of some elements are known as 'positron emitters'.* The positron did not pre-exist within that atom any more than a bark exists inside a dog; it was the energy release that created it.

The positron flies away from the atom and lives only so long as it avoids meeting an electron. As our world is made of atoms, which all contain electrons, the positron soon bumps into one, these counterbalanced opposites disappearing in a flash of gamma rays, which is light far beyond the part of the spectrum that our eyes can see. Special instruments however can detect these rays, which are exploited in medicine in the PET scanner—positron emission tomography.† Antimatter destroys, but in controlled circumstances this can paradoxically be a life saver.

On a larger scale, nature produces positrons in the heart of the sun. The sunlight that shines on us today is in part a result of positrons that were created in the centre of the sun some 100,000 years ago, only to be annihilated almost immediately.

The sun is mostly hydrogen, the simplest element. In its centre where the temperature exceeds 10 million degrees, the hydrogen atoms are disrupted into their component pieces, electrons and protons swarming independently and at random. The protons occasionally bump into one another and through a sequence of processes link together, eventually forming the seed of helium, which is the next simplest element. Helium is the ash from this fusion reaction and has less mass than the protons that were used to make it. This loss in mass has turned into energy, $E = mc^2$ at

---

* The energy materializing as a particle of matter and of antimatter—the positron.
† If you have ever had a PET scan, you will have ingested antimatter!

work, which is ultimately the energy that emerges as sunlight. So what do positrons have to do with this? A helium nucleus contains two protons and two neutrons. Under suitable circumstances a proton can change into a neutron and emit energy some of which materializes as a positron, similar to what happens in the positron emitters of earthly medicine.

The positron finds itself in the heart of the sun, where there are lots of electrons, and is instantly destroyed, turned into gamma rays. These try to rush away at the speed of light but are interrupted by the crowd of electrically charged particles, electrons, and protons that form the seething star. Buffeted this way and that, repeatedly absorbed by electrons and then emitted with less energy than before, it will take a hundred thousand years before gamma rays manage to reach the surface, hundreds of thousands of kilometres above. In doing so the rays lose lots of energy, their character changing from X-rays to ultra-violet and at last into the rainbow of colours that are visible to our eyes. So daylight is the result of antimatter being produced in the heart of the sun and, in part, of its annihilation.

This is not just a story of antimatter in history; the fusion processes that power the sun are producing positrons as you read this, and they are being annihilated faster than you can reach the end of this sentence. The gamma rays that were made just now are already wending their way upwards, eventually to emerge and illuminate the earth a thousand centuries from now.

As we shall see in our story, antimatter in the form of positrons is more common than many realize. It is put to use in medicine, technology, and science. It has been sped to within 50 metres per hour of nature's ultimate speed limit, that of light. It has also been focused into beams that are steered by electric and magnetic fields, and then smashed into beams of matter, the resulting flash of energy reproducing in a small volume for a brief moment the conditions that the whole universe would have experienced in the first moment of the Big Bang. So antimatter is enabling us to learn about the big questions of where everything has come from.

It would require more than a billion atoms in a chemical explosive to produce as much energy as could be liberated by the annihilation of a single electron. Annihilate a single gram of antimatter, (about 1/25th of an ounce), and you would obtain as much energy as you could get from the fuel tanks of two dozen conventional space shuttles. Positron energy conversion would be a revolutionary energy source which would interest those who wage war as just half a gram explosively equates to 20 kilotons, the size of the bomb at Hiroshima.[4]

It is no surprise then that if antimatter can be produced and stored until needed, it has the potential for power that would interest the space industry, or for weapons that would excite the military. I have no doubt that these possibilities are being actively investigated. This book will tell the story of antimatter, what it is, how it was discovered, how we can make it, and what opportunities and threats it could pose. It will also assess the reality of antimatter as fuel for space odysseys and for weapons.

THE MATERIAL WORLD

If you were to see a lump of antimatter, you wouldn't know it; to all outward appearances it looks no different to ordinary stuff. So perfectly disguised that it is seemingly one of the family, its ability to destroy whatever it touches would make it the perfect 'enemy within'. So, what is antimatter? Saying that it is the opposite of matter is easy on the ear, but what actually is 'opposite' about it? Knowing that the briefest contact with antimatter would commit whatever it touched to oblivion is awe-inspiring, but what gives antimatter this power?

To begin to understand antimatter, we need first to take a voyage into ordinary matter, such as ourselves. Our personal characteristics are coded in our DNA, miniature helical spirals made of complex molecules. These molecules in turn are made of atoms, which are the smallest pieces of an element—such as carbon or hydrogen or iron—that can exist and still retain the characteristics of that element.

Hydrogen atoms are the lightest of all and tend to float up to the top of the atmosphere and escape. For this reason hydrogen is relatively rare on earth, whereas in the universe at large it is the commonest element of all. Most of the hydrogen was made soon after the Big Bang and is nearly fourteen billion years old.

Vast balls of hydrogen burst into light as stars, such as our sun. It is in the stars that the full variety of elements is fashioned. Nearly all of the atoms of oxygen that you breathe, and of the carbon in your skin or the ink on this page, were made in stars about five billion years ago when the earth was first forming. So we are all stardust or, if you are less romantic, nuclear waste, for stars are nuclear furnaces with hydrogen as their primary fuel, starlight their energy output and assorted elements their 'ash' or waste products.

To give some idea of how small atoms are, look at the dot at the end of this sentence; it contains some 100 billion atoms of carbon, a number far larger than all humans who have ever lived. To see any of those individual atoms with the naked eye you would need to magnify the dot to be 100 metres across.

Elemental carbon atoms can bind in different forms, such as diamond, graphite, and carbon black—soot, charcoal, and coal. Antimatter also consists of molecules and atoms. Atoms of anti-carbon would make antidiamond as beautiful and hard as the diamond we know. Antisoot would be as black as soot, and the full stops in an antibook the same as those you see here. They too would need enlarging to 100 metres size for their anticarbon atoms to be seen. Were we able to do that, we would find that these smallest grains of anticarbon are indistinguishable from those of carbon. So even at the basic level of atoms, matter and antimatter look the same: the source of their contrast is buried deeper still.

Atoms are very small, but they are not the smallest things. It is upon entering them and encountering the basic seeds from which they are made that the profound duality between matter and antimatter is disclosed.

Each atom contains a labyrinth of inner structure. At the centre is a dense compact nucleus, which accounts for all but a trifle of the atom's mass. While enlargement of our ink-dot to 100 metres is sufficient to see an atom, you would need to enlarge it to 10,000 kilometres, as big as the earth from pole to pole, if you wanted to see the atomic nucleus. The same is true for antidots and anti-atoms. It is only when they are seen in such fine detail that the subtle choice of matter or antimatter begins to show.

When the profound entangling of space and time that comes with Einstein's theory of relativity is married with the will-o'-the-wisp ephemeral world of uncertainty that rules within atoms, an astonishing implication emerges: it is impossible for nature to work with only the basic seeds of matter that we know. To every variety of subatomic particle, nature is forced also to admit a negative image, a mirror opposite, each of which follows the same strict laws as do conventional particles. As the familiar particles build atoms and matter, so can these contrary versions make structures that at first sight appear to be the same as normal matter, but are fundamentally dissimilar.

Matter and Antimatter

Inside atoms we find swirling electric currents, powerful magnetic fields, and electrical forces that attract some things and repel others. Within atoms of antimatter these currents, fields, and forces are also present, but their polarities are reversed: north poles become south; positive charges become negative. Imagine our ink-dot and antidot enlarged to 100 metres so that we can see their individual atoms or anti-atoms. Gently propel a tiny magnet towards the outlying regions of an atom, then launch it at an anti-atom and compare what happens. A gentle curving to the left for the one case becomes a mirror image arc to the right for the other; where once it was pulled in, now it is pushed out; where previously it was rejected, now it is sucked in; where before it was safe, now it faces annihilation.

The source of these forces is the atomic nucleus, which is electrically charged. As magnets have north and south poles, giving them the power to attract or repel one another, so the rule of electric charge is that like charges repel and opposite charges attract. In normal matter the atomic nucleus carries positive charge; electrons, the tiny lightweight particles that occur in the outer reaches of atoms, are negatively charged. An atom of the simplest element, hydrogen, consists of a single electron remotely encircling the central nucleus, which consists of a single proton.

It is the mutual attraction of opposite electric charges that keeps the negatively charged electrons gyrating remotely around the central positively charged nucleus. It is these electric and magnetic forces at work deep within atoms that provide the tentacles by which molecules and macroscopic structures—crystals, tissues, rocks, and creatures—are organized and held together.

The force of gravity rules the galaxies, planets, and falling apples, and keeps our feet on the ground. However it is electric and magnetic forces that give us shape and structure. The electromagnetic force is much more powerful than gravity, but in bulk matter the attractions and repulsions of the positive and negative charges tend to cancel out, leaving the all-attractive force of gravity as dominant. Thus although intense electrical forces are at work deep within the atoms of our body, we are not much aware of them nor are we ourselves electrically charged.

Nonetheless, there are plenty of clues to this inner structure, which is revealed in situations where the effects of the positive and negative charges do not precisely cancel. A build up of unbalanced electric charge causes sparks, as in lightning; a magnet can attract a lump of metal, overcoming the downward gravitation of the whole earth. On a larger scale, swirling electric charges in the earth's core make the entire planet a huge magnet, which is revealed when a small compass needle swings in line with the earth's magnetic field, pointing to the north and south magnetic poles.

That is what was known in 1928 when the story of antimatter began. Atoms, as understood by Paul Dirac, Carl Anderson, and Robert Millikan, the principal players in the first act of the antimatter saga, consisted of dense clumps of massive protons, whose positive electrical charge trapped negatively charged lightweight electrons in a cosmic waltz.* Armed with this knowledge, we can begin to appreciate the idea of antimatter.

The laws of electricity and magnetism that underlie the existence of bulk matter don't care which bits of matter carry negative charge, and which bits are positive. If we swapped all positives to negative, and all negatives to positive in some situation, the resulting forces would be the same and the structures they built would also be unchanged. If one imagined that all negatively charged electrons were instead positive, and in compensation all

---

* The rules of electricity would lead us to expect that a crowd of protons would fly apart from their mutual repulsion. However, experience and experiment shows another more powerful force at play, felt by protons but not electrons, which binds protons into tight bundles forming atomic nuclei.

protons were negative, to all outward appearances nothing would appear different.

Such a swapping of charges would turn what we know as matter into what we call antimatter. An anti-atom of antihydrogen would consist of a negative 'antiproton' encircled by a positively charged 'positron'. Paul Dirac, who first predicted that such a mirror image of matter should exist, summarized this enigma on receiving his Nobel Prize in 1933:

We must regard it rather as an accident that the Earth (and presumably the whole Solar System) contains a preponderance of negative electrons and positive protons. It is quite possible that for some of the stars it is the other way about, these stars being built up mainly of [positively charged electrons] and negative protons.

With great prescience, and fully appreciative of the deep symmetry between the positive and negative, he commented that half of the stars could be of one kind, and half the other. These are what today we would call matter and antimatter, and as we look into the night sky at those stars, there would be no way of distinguishing them.

Spectra and the Quantum Electron

It is only within the subatomic universe that these two contrary forms of substance are revealed. This is a land whose laws appear bizarre to our experiences in the world at large. It would be in attempting to understand the implications of those laws that science stumbled upon the inevitability of antimatter.

Isaac Newton's laws of motion, which govern the behaviour of visible things where countless billions of atoms act in concert, predict with certainty how billiard balls will bounce. Things are very different for individual atoms and their constituent particles, which occupy a world of indeterminacy where only the relative chances of things happening can be predicted. Whereas billiard balls bounce from one another in a determined way, beams of atoms will scatter in some directions more than others, forming areas of intensity or scarcity like the peaks and troughs of water waves that have diffracted through an opening.

The behaviour of individual atoms may appear random, but in reality it is not. Atoms are described by the laws of 'quantum mechanics', which predict the probability that a particular atom will do this or that. Just as I cannot with certainty predict if the toss of an individual coin will yield a head or a tail, nonetheless if I toss millions of them, I can be certain that the ratio of heads to tails will be nearly one, and the more tosses that are made, the more certain I will become. So it is with atoms. The fundamental laws of quantum mechanics apply to each individual atom; I cannot with certainty predict how an individual atom will respond when it's hit, whether it will metaphorically land head or tails, but when millions of them are involved, the random chances of head and tail gradually even out. When large numbers of atoms are involved, Newton's laws of certainty emerge from the underlying quantum rules.

Newton's laws predict that the motion of balls made of matter would be identical to those of antimatter: billions of atoms behave the same as would billions of anti-atoms. However it is within individual atoms that the bipolar nature of matter lurks, and it is there that the quantum laws rule. It is these quantum laws, when combined with Einstein's theory of relativity, that reveal that just one form of matter is not enough: the act of creation in the Big Bang must have made two counterbalanced varieties.

In popular accounts, atoms are often described as miniature solar systems, with the planetary electrons whirling around the nuclear sun: little things whizzing around something big in the middle. However, as soon as this picture was first proposed, people worried about it.

The earth takes a year to orbit the sun, and has done so for over four billion years without harm. Contrast this with the electron in a hydrogen atom, which is apparently orbiting around the central proton at about one per cent of the speed of light, making some million billion circuits each second. Said another way: in a millionth of a second an electron makes more circuits of the central proton than the earth has made around the sun in its entire history. According to the theory that existed at the start of the 20th century when these ideas began to emerge, such an electron would be emitting so much electromagnetic radiation that it should have immediately spiralled into the nucleus in a blaze of light. So how do atoms survive; how can anything exist?

It was quantum theory that provided the answer. When you get down to distances smaller than a millionth of a millimetre, which is the scale of atoms, our experience of everyday things is a poor guide about what to expect.

In 1900 Max Planck had shown that lightwaves are emitted in distinct microscopic 'packets' or 'quanta' of energy known as photons, and in 1905 Einstein showed that light remains in these packets as it travels across space. This was the beginning of quantum theory, the idea that particles can have will-o'-the-wisp properties, being neither here nor there but 'most likely here, possibly there'. In quantum mechanics, certainty is replaced by probability, which peaks and falls like a wave. Its immediate success was in explaining how atoms could survive.

The quantum waves of chance can be imagined as waves on a length of rope. If the rope were coiled in a circle like a lasso, the length of any wave would have to fit neatly into its circumference. Think of the circular loop like a clock face. If there is a peak at twelve o'clock and a dip at six o'clock, the next peak fits perfectly at twelve. However a wave peaking at twelve with a dip at five o'clock would have its next peak at ten, leaving the twelve o'clock point out of time with the beat of the wave. The Danish physicist Niels Bohr realized in the summer of 1912 that the waves of chance for electrons circulating in atoms also must fit perfectly into each loop; electrons cannot go anywhere they please but only on those paths where their waves fit precisely. In particular they cannot spiral into nuclear destruction: the atom is stable. (See figure 1.)

These quantum waves also explained a mystery that was two centuries old: the phenomenon of atomic spectra. It is relatively easy to shake light out of atoms and have them reveal their unique spectra. You can do so by adding some element such as sodium to a flame and looking at the light through a prism or a diffraction grating, which splits it into its component colours. These will include a series of bright lines, which in the case of sodium include two particularly intense yellowy-orange ones, which are the source of the familiar colour of sodium street lamps. Analogously mercury vapour lamps glow a bluey-green, while the pink light that permeates many pictures of the stars is due to hydrogen's tendency to emit some visible light at the far red end of the rainbow. These beautiful coloured patterns demanded explanation: what causes them? Why do they vary from one element to another? We now know that they are the result of the quantum motions of electrons within atoms.

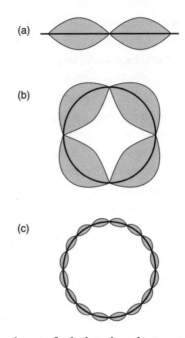

Figure 1. Waves have to fit the loop in order to survive.

Light is only emitted when an electron moves from one path to another. If the initial path only accepted electrons with high energy, and the electron switched to a path where the energy is lower, the difference between the two energies is taken up by the photon of light that is emitted. The total energy remains the same, it just gets redistributed. Thus these photons can only have certain discrete amounts of energy, as determined by the particular jumps that the electron can make. The discrete values of the photons' energies are seen by our eyes as differing colours. The result is that the emitted light gives the spectrum of colours that is unique to each atomic element. It is from these colourful autographs that it is possible to tell which atomic elements are present throughout the cosmos as they beam their spectra at us. These coloured patterns are the visible proof of the quantum waves of ordered chance that rule in the subatomic world of fundamental particles.

The Spinning Electron

It will be the electron that heralds the realm of antimatter, the mysterious shadow to our material world. Ever since J. J. Thomson in 1897 first produced the electron in isolation, liberated from its atomic prison, we have known that electrons exist and that their presence within atoms is the source of spectra. Even before Thomson had proved this beyond doubt, scientists already suspected that this atomic constituent exists, and had even deduced that it had both electric charge and a twofold magnetism, akin to the north pole–south pole duality of a familiar bar magnet. Half a century later this would be explained by Paul Dirac and lead him to predict the existence of antimatter.

In 1896 Peter Zeeman, the Dutch spectroscopist, had noticed that when powerful magnets were near his samples, the bright yellow lines emitted by sodium subtly changed. These lines are normally very sharp and well defined, but Zeeman noticed that they broadened in a magnetic field. Later, more powerful instruments were developed which showed that what had appeared to be a broadening was actually a separation of one line into two or more. The separation of these individual lines was too small to be seen by Zeeman, and had appeared as a broad fuzz, as if viewed by a short-sighted person without spectacles.

It turned out that this was because the electron has magnetism. As two magnets can attract or repel depending on how their north and south poles are arranged, so does the electron's motion in a

25

magnetic field affect its energy. In consequence this will slightly modify the energies of any photons that are emitted, and hence alter the pattern of the spectral lines.

The 'Zeeman effect' revealed that an electron can act like a little magnet with a north and a south magnetic pole of its own. It was as if the electron has an intrinsic rotary motion, known as 'spin', which can take on either of two orientations in a magnetic field: clockwise or anticlockwise if you like. The idea that an electron with no measurable size can 'spin' makes no sense in everyday terms, but it is a word that physicists freely use when referring to this bizarre property. The hypothesis that the electron has this duality certainly explained a wealth of data in atomic spectroscopy, but for many years the idea of 'spin' was little more than a desperate attempt to make sense of lots of data. Where this property originated; why it occurs; these were mysteries that would only be explained when Dirac combined relativity and quantum mechanics.

E *is for Einstein and* E $= \mathrm{mc}^2$

Spin and antimatter both emerge as necessary properties of the physical world when the quantum laws and Einstein's theory of relativity are joined together. It was Einstein who first showed what energy really is, with the astonishing consequence that matter is trapped energy. When energy congeals into particles of matter it produces a negative imprint, which is antimatter. It would be Paul Dirac who first discovered that profound truth.

The classical laws of motion were discovered by Isaac Newton over 300 years ago. First there is his law of inertia: bodies are 'lazy' and stay at rest or at constant speed unless something forces a change. Bodies have a reluctance to be shifted from their stupor or steady movement: experience shows that it is easier to move a leaf than a lump of lead. Newton asserted that if the same amount of force was applied to two bodies, their relative acceleration would be a measure of their intrinsic inertia, or mass.

An immovable object, the central character of my father's first conundrum, would have to have infinite mass. Such a concept is impossible, at least in Newton's mechanics, as all the mass in the universe, though indeed huge, is not infinite. However, since Albert Einstein rewrote our world-view in his theory of relativity, where space rolls up and time distorts, the idea of something having an infinite mass and becoming utterly resistant to acceleration has become a reality.

If some body is stationary and you apply a force to it for a second, its speed will increase by some amount, say 10 metres per second. Now apply that same amount of force again. According to Newton, and everyday experience, the speed will again increase by 10 metres per second. If you repeat this, the body will get faster and faster without limit. However, Einstein says that if you were to measure the change in speed very precisely you would discover that although it rose by 10 metres per second when forced from rest, the next push will speed it by fractionally less than 10 metres per second, and as it moves faster, it will become ever harder to accelerate. Were it moving at near to the speed of light, the application of the force would hardly alter its speed at all.

Newton's rules are an excellent approximation to the exact laws of motion so long as we deal only with objects that are moving slowly relative to the speed of light. As light rushes by at 300,000 kilometres per second, Newton's rules are very accurate in day to day affairs; however, if you are concerned with the behaviour of electrons in a particle accelerator where speeds can get to within a fraction of a per cent of light speed, then Einstein's more complete description has to be used.

In Einstein's theory of relativity the mass of a body gets larger and larger the faster it travels. As it approaches the speed of light, the mass grows extremely fast, making the object ever more resistant to acceleration. Eventually, as one tries to reach the speed of light, the mass becomes infinite. It is thus impossible to accelerate a massive object to the speed of light; the only things that travel at light speed are things with no mass, such as light itself!

Although the idea that inertia changes with speed may seem peculiar to our 'common sense', it is nonetheless true as years of experience with high energy particles shows. As particles of matter rush around the racetrack at laboratories such as CERN, to meet beams of antimatter hurtling in the opposite direction, the timing is critical and relativity has to be invoked in order for them to arrive on cue.

One of the consequences is that the relationship between energy and motion that had been known since Newton, and had been assumed by the pioneers of the new quantum mechanics, with initial success in describing the behaviour of atoms and electrons, is actually more subtle.

A surprising and far reaching implication of Einstein's theory of relativity is that even a stationary object contains energy, which is locked within its constituent atoms. The amount of this energy is the 'E' in his famous equation $E = mc^2$, where m is the amount of mass and c is the speed of light. It is latent within matter, even when it is standing still.

To keep the total energy accounts for a moving body, its kinetic energy has to be included in the sums. The natural guess is that you simply add the kinetic energy to the energy contained in its mass (mc^2). This would be true but for the fact that when in motion the object's mass m increases and so the magnitude of mc^2 also changes. Although working it all out was a tricky business, the answer for the total energy E of a moving body turned out to be rather simple. You calculate it by first adding the *square* of the energy of motion to the *square* of the energy in its mass mc^2; having done so, the square root of this will be the answer. So for example if the amount of energy at rest was four joules, and the motion gave a further three joules, the total would be five joules (as three threes, added to four fours, gives a total of twenty-five, which is the same as five fives).

A pictorial representation of this sum (see also figure 2) is to draw a right-angled triangle whose sides have lengths proportional to various amounts of energy. The base represents the energy contained in its mass, mc^2. The length of the vertical is then proportional to the kinetic energy.* The length of the hypotenuse is then proportional to the total energy of the object. Remembering the old rule that 'the square on the hypotenuse equals the sum of the squares on the other two sides', we have the simple accounting for the total energy E of a massive body in motion: the total amount of energy, all squared, equals the sum of the amount of energy mc^2 contained in its mass, all squared, added to the energy of its motion, squared.

The implications of Einstein's theory of relativity for the nature of energy are astonishing. First, massive objects at rest contain

---

* If you want more mathematical precision here, it actually represents the amount of energy given by its momentum multiplied by the velocity of light: it is usual to refer to the momentum by the symbol p and the speed of light by c, hence the product is pc. The length of the hypotenuse is then the total energy of the object, given by $E^2 = (mc^2)^2 + (pc)^2$.

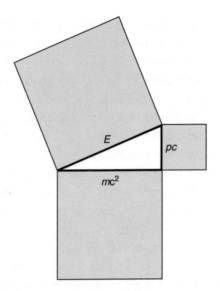

Figure 2. Einstein, Energy, and Pythagoras' theorem. In Einstein's theory of relativity the amount of energy E of a body in motion is proportional to the area of the square on the hypotenuse of a right-angled triangle whose base is proportional to the magnitude of its energy when at rest (mc^2), and its energy of motion, proportional to the product of its momentum and speed of light, pc.

an amount mc^2 of energy trapped within them. Second, even something that has no mass, such as a photon travelling at the speed of light, will have energy due to its motion. As energy overall is conserved, it is therefore possible for the energy in a beam of light to be transformed into energy trapped within matter.

But how can an electron with negative electrical charge emerge from the energy in a puff of light, which has no electric charge? This is where nature's two forms of matter enter the story. The negatively charged electron has a positively charged form known as the positron. The energy of a photon, a particle of light, becomes trapped in these two complementary pieces of substance. This process can also happen in reverse: an electron and a positron can annihilate one another, their individual energies being taken by the photons that rush from the scene of destruction at the speed of light.

The emergence of substance from pure energy, of which the purest form is light, is almost biblical in scope. With antimatter, the negative image of matter, we make contact with the gods of creation. Here we begin to see how our universe emerged from the Big Bang. Intense heat and light with huge energy congealed into counterbalanced pieces of matter and antimatter. Einstein's theory of relativity, with its profound implications for the nature of energy, suggests how matter was created at the start of time. An essential part of this is the idea that matter has a mirror image: antimatter. While relativity explains the energy accounts, it is when relativity is combined with quantum mechanics that the full power of nature is revealed. It was from this union of the two great theories of the 20th century that the idea of antimatter would be born.

3

TABLETS OF STONE

Paul Dirac

As a teenager, and before I became focused on science, I was an avid reader of anything that the library in Peterborough had to offer. Upon borrowing a book, which you could have for up to two weeks, the librarian would stamp the return date on a sheet stuck inside the cover. Some books were so popular that several sheets were already saturated with date stamps and, in those pre-Amazon.com days, this was the best record of what was likely to be a good read. This was a criterion for deciding what to borrow, but I began to tire of reading what everyone else had read and wondered which books had been read the least. Many had only been loaned once or twice, usually because they were recent purchases, but at last I found a book that had been chosen only once, and that had been many years earlier.

I became the second borrower, although as it turned out I read only the first paragraphs of the preface. It kept mentioning 'orthogonality', which I didn't understand, neither could my father explain what the book was about. This was one of those memorable points in life—the first time that one's parents are utterly stumped; clearly this book was something truly special.

A decade later, by which time I was in my final year as an undergraduate studying theoretical physics, I came across the book again. This time I was able to follow most of it, but even then only with considerable difficulty. It was *The Principles of Quantum Mechanics* by the Cambridge mathematician, Paul Dirac, and I discovered that not merely my father and I, but almost everyone else had had trouble following the arguments in the original version, published in 1930. It is the second edition, which was completely rewritten and published in 1935, that for over seventy years has been the classic textbook for any serious student of quantum mechanics, although even this is hardly an easy read. The book is a paradigm of concise logic, full of mathematical equations, interspersed with explanatory text, in which Dirac describes his unique and revolutionary approach to physics, including his eponymous equation that predicted the existence of the positron, the simplest particle of antimatter.*

The first glimpse of the antiworld came not from experiment, a chance discovery, but from the beautiful patterns that Dirac had seen in his equations. As crotchets, minims, and semiquavers on a stave are mere symbols until interpreted by a maestro and

---

* During the next vacation I went back to the Peterborough library and found the version that I had borrowed ten years before. Having verified that it was indeed the same book, I looked inside the front cover at the date stamps. There were only two, one being mine; I never did discover who else it was that had borrowed it all those years before.

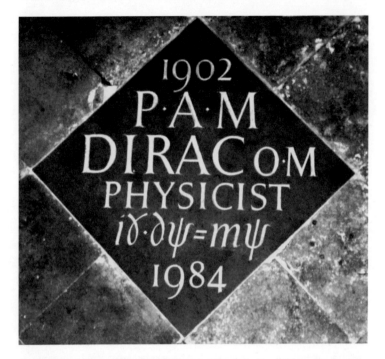

Figure 3. Memorial to Paul Dirac in Westminster Abbey displaying his equation. The symbol γ refers to any of the 'gamma' matrices described in Appendix 2: 'The Dirac Code'.

transformed into sublime melody, so can arid equations miraculously reveal harmony in nature. It was in the language of mathematics that Dirac was a supreme master. When in 1995 a plaque in his memory was unveiled in Westminster Abbey (see figure 3), adjacent to the memorial to that greatest of scientists, Isaac Newton, it was seen to display his famous equation, the one that revealed the antiworld. It has a raw beauty, even to the vast majority of visitors who do not know what the symbols mean. To those who have learned to read the hieroglyphs of mathematics,

the creativity, power, and elegance of Dirac's equation compare
with Shakespeare or Beethoven.

$$i\gamma \cdot \partial\psi = m\psi$$

Paul Dirac's father was Swiss, and had moved to Bristol where
he taught languages. French and English had equal place in the
Dirac household and Paul was brought up bilingual, although he
was unusually taciturn in both languages. Stories of his linguistic
economy and lack of communication skills are legion. This con-
trasted with his mathematical abilities, which were immense. His
lectures, which were mathematically brilliant and semantically
precise, could be daunting even for experts. During a lecture at the
University of Toronto a member of the audience asked politely 'I
do not understand how you derived that formula on the board'.
There was a long silence, and it was only after being prompted
by the chairman to give an answer that Dirac responded 'It was
not a question; it was a statement'.[1] At college dinners there was
always the delicate issue of who would have the mixed privilege
of sitting by the silent mathematician. On one occasion, when
novelist E. M. Forster was a guest, the college had the inspiration
of setting the pair together; Forster also was more at ease with
written words than conversation and Dirac was an avid reader of
Forster's works. According to folklore, which is probably apoc-
ryphal but could well be true given the characters, the evening
developed as follows.

Through the soup course nothing was said but as the main
dish was served, Dirac leaned over and, in a reference to Forster's
Passage to India, asked 'What happened in the cave?' That was to
be Dirac's contribution to the evening. Forster pondered Dirac's
question but sat silent. He ate on and ruminated further. Finally
the desert arrived and Forster delivered his answer: 'I don't know'.

35

Limited though Dirac was in personal communication, like Forster he expressed himself through written symbols. In his great work of 1928, where he brought together the ideas on the quantum theory and fused them with the other magnum opus of the century, Einstein's theory of special relativity, Dirac invented a whole new mathematical language. Strange and bizarre it appeared at the time, but today it is part of the standard education of students of theoretical physics and used by all practitioners of the subject.

Two for the Price of One

Mechanics is the science of motion. It describes how things move from one point to another as time passes, the greater the distance moved each second so the greater is the speed. If something moving hits you, the impact will depend not just on how fast it's travelling but also how massive it is. It is the momentum that matters: the product of mass and velocity. Mechanics also deals with energy, especially the energy due to motion, 'kinetic energy'. In everyday experience this grows in proportion to the square of the speed: that is why fast tennis serves are so hard to do—to double the speed you have to give four times as much energy to the ball.

You cannot know with precision both the position of a particle and its momentum, but the amount of uncertainty is so trifling as to be unmeasurable when dealing with objects that are large enough to see, made of billions and billions of atoms. However, for things that are very small, such as atoms and their constituent particles, this 'unknowability' becomes overwhelming. The basic rules of normal mechanics had to be carefully rewritten to take this uncertainty into account; the result is what we call Quantum Mechanics.

In common with normal mechanics, the equations of quantum mechanics deal with energy, momentum, time, and position just as before. So long as you know how the energy, mass, and momentum of a particle are related, the quantum equations will enable you to calculate what will happen in any event. The challenge is to define that relation.

The notion of the quantum, that lightwaves act as particles called photons and that particles such as electrons have wavelike characters, had emerged early in the 20th century. However, a quarter of a century passed before the equations of the quantum mechanics were discovered.

Erwin Schrodinger in 1926 solved this for the case of slow-moving particles, 'slow-moving' meaning relative to the speed of light. The 'Schrodinger Equation' explained the behaviour of electrons in atoms, and showed that in a hydrogen atom the electron is effectively moving with a speed of about two thousand kilometres a second. This is fast to our senses but is less than one percent of the speed of light. Schrodinger's theory worked, and even today is widely applied to problems in atomic physics.

Schrodinger's equation also explained why the orbital motion of electrons in atoms caused the spectral lines to multiply in magnetic fields. However, it gave no explanation for the electron's own intrinsic 'spin'. This known property of the electron had no place in Schrodinger's theory. A more complete quantum mechanics, one that incorporated spin and applied at relativistic speeds, waited to be discovered.

The challenge begins with the subtle nature of energy in Einstein's theory of relativity. Recall that a massive body contains energy ($E = mc^2$) trapped within its constituent atoms even when

37

it is at rest, and when moving it has kinetic energy also. As we saw on page 30, the total energy is given by an analogue of Pythagoras' rule that for a right angled triangle 'the square on the hypotenuse equals the sum of the squares on the other two sides'. For a massive body in motion, the total amount of energy, all squared (E^2), equals the sum of the amount of energy when at rest, all squared, added to the energy of its motion, squared.

Oscar Klein had tried to generalize Schrodinger's theory by using E^2 and Einstein's 'hypotenuse' relation. As the square root of 25 can be either +5 or −5, so the square root of E^2 could be either positive or negative. Because the hypotenuse of a triangle has a positive length and not a negative one, the negative solution for the energy that Einstein's Pythagorean relation seemed to allow was treated as spurious. Even so, it left people uneasy. The problem arose because the original equation was written for the amount of energy *squared*. Dirac decided to avoid this conundrum by restricting himself at the outset to E rather than E^2.

This was a natural thing to attempt but not so easy to do. The problem facing him was how to write a relation between the length of the hypotenuse (energy E), and the lengths of the other two sides (the energy when at rest, mc^2, and kinetic energy), each of these appearing just once rather than squared. It was in doing this as his basis for a quantum mechanics that everything fell into place.

To get the accounting right, Dirac needed to find two quantities that when multiplied together give zero, whereas each individually squared gives one. This is obviously impossible. The product being zero requires one of the numbers to be zero, and so its square will be zero, not one.

At this point, if not before, most would give up, convinced that the task is impossible. There is a clever way that it can be done, and Dirac discovered how. If you are interested in the mathematical trick that he used to crack the problem, see the Appendix 2, 'The Dirac Code', on page 152.

Dirac had realized immediately that it was impossible if the two quantities are simply numbers, but that it can be done if they were 'two dimensional numbers' known as matrices: an array of two columns with two numbers in each column. Mathematicians have worked out the rules for adding and multiplying matrices and they are used to keep the accounts in many cases in engineering, electricity, and magnetism. They have one intriguing property that is the key to solving Dirac's enigma: if you multiply two matrices in the order $a \times b$ the answer is not necessarily the same as if you do it in reverse order, $b \times a$. This might at first sight seem odd, but there are many examples of things where the order matters.

Anyone who has played with Rubik's cube knows that twisting the top clockwise and then rotating the right hand side to the back gives a different pattern than if you did the two operations in the reverse order. It is easier to see this with a die (see figure 4). If you rotate a die clockwise and then about the vertical, it will be oriented differently to the case where you had first rotated about the vertical and then clockwise. This is why matrices have proved so useful in keeping track of what happens when things rotate in three dimensions, as the order matters.

So if the two quantities a and b that Dirac was looking for are matrices, they can solve his problem. They can satisfy $a^2 = 1$; $b^2 = 1$. And although $a \times b$ is not zero, neiher is $b \times a$ zero, their sum can be: $a \times b + b \times a = 0$. Using matrices, Dirac was able to write an equation relating the total energy of a body to a sum of its energy at rest and its energy in motion, all consistent with Einstein's theory of relativity.

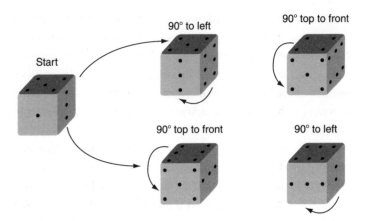

Figure 4. A die turned 90 degrees to the left and then 90 degrees from the top to the front, reveals a different face than when the order is reversed.

The fact that matrices keep account of what happens when things rotate was a bonus, as the maths was apparently saying that an electron can itself rotate: can spin! Furthermore, the fact that he had been able to solve the mathematics by using the simplest matrices, where a single number was replaced by two columns of pairs, implied a 'two-ness' to the spin, precisely what the Zeeman effect had implied (page 25). The missing ingredient in Schrodinger's theory had miraculously emerged from the mathematics of matrices, which had been forced on Dirac by the requirements of Einstein's theory of relativity.

This was remarkable in itself, but there was a further feature that tantalized him. All worked well so long as he regarded solutions with positive energy *and* negative energy as real. In trying to avoid the negative energy problem that had arisen when others had worked with E^2, Dirac had been forced to use matrices and had explained the electron spin, but with the irony that the negative energy solutions insisted on being treated as legitimately as the positive ones. Putting these two sets of solutions together meant that he had two sets of 'two-by-two' matrices. In effect, he had been forced to write his theory with simple numbers replaced by matrices that consisted of four columns with four numbers in each.

These four-column matrices are known as the 'γ' (gamma) matrices, and that is the origin of one of the symbols in the eponymous equation engraved in Westminster Abbey. Apparently Einstein could only be satisfied if an electron both had spin and also either positive or negative energy. Dirac had set out in the hope of avoiding this enigma of negative energy, but had been forced to accept it. What could this mean?

The Infinite Sea

When you press the accelerator pedal in a car, the car speeds up: it gains energy of motion, 'kinetic energy'. This hasn't come from nothing, it has involved burning fuel, which has released energy from within it and turned it into an equal amount of kinetic energy of the car. Press the brakes and the car slows; its kinetic energy falls. This energy hasn't disappeared but has converted into heat in the brakes and tyres, and possibly sound if you skid. Eventually, you come to rest. Your kinetic energy is zero but there is still much energy potentially available locked into your petrol tank. Even if your tank is empty there is a vast amount frozen into the mc^2 of the atoms that make you and the car, and you could use some of your own mc^2 to give the car kinetic energy by the simple expediency of pushing it.

41

The idea of trading one variety of positive energy for another is what powers industrial society. There is no sign of negative energy in these day to day experiences, so what do the negative energy solutions for an electron mean?

If electrons could have negative energy, then one would expect that electrons in matter could spontaneously lower their energy by dropping into one of the negative energy states. As this would make matter unstable, the fact that we are here at all seemed to imply that Dirac's theory of the electron must be wrong, and that there cannot be any such negative energy possibility. Remarkably Dirac used the fact that matter *is* stable to interpret the negative energy states! To begin to understand how, we need to appreciate a remarkable regularity among the atomic elements, discovered by the Russian Dmitri Mendeleev and encoded in his 'Periodic Table'.

Some elements are quite similar in their properties, these similarities recurring 'periodically' when one lists the elements in the order of increasing atomic masses. Examples of properties that recur periodically are the chemical inertness of the gases helium, neon, and argon; metals that have affinity for water such as sodium and magnesium; and the highly reactive elements fluorine, chlorine, and iodine whose affinity for hydrogen makes acids. These similarities have been known for centuries. Mendeleev's Periodic Table revealed their periodicity, but it was quantum mechanics that explained them, and does so in a way that solves Dirac's dilemma.

The electrons in all atoms are identical. The difference between one variety of atomic element and another is the number of electrons orbiting the central nucleus (and of course the number of protons in that nucleus). As we have seen, these electrons cannot go anywhere they please, instead the laws of quantum mechanics restrict them to a few specific paths or 'quantum states'. The patterns of available paths as you add more electrons, and hence move through the table of elements, keep recurring in a regular cycle such that the periodic similarity of the ensuing elements also recurs. Quantum theory explains this as a consequence of a fundamental rule known as the exclusion principle. In effect, electrons are like cuckoos, where two in the same nest is one too many, or in the drier language of quantum mechanics: no two electrons in some collection can occupy the same quantum state.

Having realized that his equation implied that electrons could have negative energy, Dirac used this exclusion principle as the basis of a brilliant idea. He suggested that what we call the vacuum is not actually empty but is like a bottomless pit, descending into which is a ladder each of whose rungs corresponds to a possible quantum state, a resting place for an electron. The top of the ladder corresponds to zero energy, all the rungs below being the possible negative energy states for electrons. Dirac's insight was that if all of these negative levels are already filled, no electrons can fall into a negative energy slot and so matter remains stable. What we call the 'vacuum' would be like a deep calm sea that is unnoticeable so long as nothing disturbs it. The filled sea is the base level relative to which all energies are defined: Dirac's 'sealevel' defines the zero of energy.

In Dirac's interpretation of the vacuum, if one electron in this sea were missing, it would leave a hole. The absence of a negatively charged electron with energy that is negative relative to sea-level, will appear as a positively charged particle with positive energy, namely with all the attributes of what was later called a positron. This was a strange idea, and quantum mechanics is still strange eighty years later; it was only in its infancy when Dirac made his proposal, which was a piece of radical genius.

43

How could a negative energy electron be dislodged so that a hole in the vacuum sea could be seen? The answer is to supply energy, for example by a high energy gamma ray. If the gamma ray had enough energy, it could kick an electron from a negative energy state into a state with positive energy. The result will be that the gamma ray has produced both a positive energy electron and also a hole in what was the vacuum. The hole is an absence of both negative energy, which will manifest as a positive energy state, and negative charge, which appears as a positive charge. So the end result is that the energy of a gamma ray has turned into a conventional negatively charged electron, accompanied by a positively charged electron, both of positive energy. (See figure 5.)

What is this Positive Electron?

Dirac's prediction of the anti-electron seemed to many at the time to be science-fiction. Up to that time the only particles known were the electron and proton, from which all matter could apparently be explained. Furthermore they were believed to be immutable, yet Dirac's theory implied that these fundamental particles of matter could be created or destroyed at will. There was no need for further particles or any desire for them, apart from the generally accepted but yet to be discovered neutron, which adds bulk to the atomic nucleus and helps stabilize it. The days when weird particles with fanciful names would proliferate as a result of discoveries in cosmic rays and particle accelerators, were still far in the future. In 1928 the particle picture was simple: matter is made from negatively charged electrons and positive protons. In this relatively cosy world-view, the anti-electron had no place.

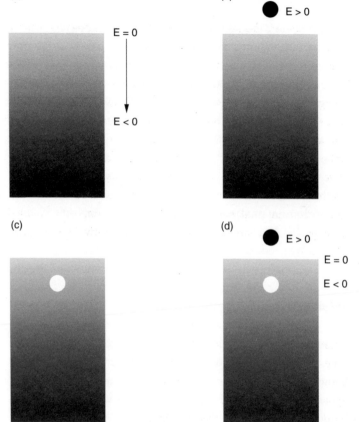

Figure 5. The vacuum is filled with an infinitely deep sea of energy levels from negative infinity to some maximum. We define this configuration, the state of lowest energy, to have zero. An electron with positive energy relative to the vacuum is the black circle. A missing state with negative energy (the white circle) and negative charge will appear as if a positive energy state with positive charge. This is Dirac's picture of the antiparticle of the electron: the positron. If a negative state is empty and a positive one also filled, this could be a positive energy electron and the 'hole' perceived as a positive energy positron. To produce this configuration, energy must first be supplied to the vacuum. This energy could be donated by a photon whereby the photon has converted into an electron and a positron.

At every talk that Dirac gave in the months following the publication of his theory, people would ask him 'where is the anti-electron?' Such questions invoked nervous laughter and Dirac soon tired of them. That few of his contemporaries had seriously studied his equation or were able to follow his arguments made this 'smart' question particularly galling. Eventually he attempted to forestall them by suggesting that as a proton has positive charge, the anti-electron, with its positive charge, might actually be the proton.

This led to a widely held opinion that Dirac actually seriously considered the proton as a candidate for the positively charged particle that had emerged from his equations like a rabbit from a magician's cloak. To modern physicists, cognizant of the profound symmetry between matter and antimatter, the idea that the proton could be identified as the anti-electron, when there is a mismatch of nearly 2,000 in their masses, seems absurd.

It all might seem obvious today, but the deep symmetries between matter and antimatter are much better understood now than in 1928 and historians argue whether the mismatch in masses was so obvious a nonsense as it now appears. Peter Kapitsa, a famous Russian physicist and contemporary of Dirac, has claimed that Dirac made the remark as a joke, his taciturn character notwithstanding. Dirac's aim was simply to quieten the persistent questioners, allowing him to get on with explaining his profound ideas and to leave the question of the mass as a 'detail' to be solved later.

Dirac's equation had pointed the way towards antimatter, but it was Robert Oppenheimer, later to become famous as leader of the Manhattan atomic bomb project, who really saw the full vision. He pointed out that the positive particle could not be the proton, for if it were, then hydrogen atoms would self-destruct. The arguments that implied an electron and its positive counterpart can emerge from the vacuum, could be applied in reverse: play the film backwards and it would show the pair mutually annihilating, disappearing into gamma rays. So if the positive particle were identified with the proton, hydrogen atoms would survive only so long as the proton did not meet the electron: not just hydrogen, but all matter would vanish in a flash of light.

Dirac immediately realized the power of Oppenheimer's criticism and accepted that his positive electron was indeed something entirely new. In September of 1931 he published his conclusion that a hole would be a 'new kind of particle, unknown to experimental physics, having the same mass and charge as an electron. We may call such a particle an anti-electron'[2]. Its dramatic property would be its pyrotechnic destruction by a conventional negatively charged electron, and the converse possibility of their mutual creation from pure energy.

The massive proton is another beast entirely and Dirac's equation implied that it too has an 'anti' counterpart. In his 1931 paper he also made this clear, writing that in his theory 'there is a complete and perfect symmetry between positive and negative electric charge'. With only a slight hint of caution he continued that 'if this symmetry is really fundamental in nature, it must be possible to reverse the charge of any kind of particle'. Thus he predicted that an antiproton, a negatively charged massive mirror of the proton, should also exist. Dirac's prediction that to every particle variety there exists an antiparticle counterpart is now recognized to be an essential truth, a glimpse of a profound symmetry in the tapestry of the universe.

Dirac's idea of the bottomless sea full of identical electrons also explains why all electrons and positrons created 'out of the vacuum' have identical properties rather than emerging with a random continuum of possibilities. Dirac also proposed that protons fill the sea, and today we recognize that their more basic seeds, the 'quarks' (which we shall meet in chapter 5), also satisfy the exclusion principle and fill an infinitely deep sea. It is the infinitely deep storehouse of the Dirac sea that provides us with the particles, and antiparticles, that we can materialize.

So much for antimatter in theory; what is the reality? The positron's story is where we go next.

A COSMIC DISCOVERY

Thousands of metres above our heads, high energy torrents of subatomic particles and gamma rays from outer space are crashing into the upper atmosphere. These collisions spawn a deluge of further particles, most of which are absorbed in the air before reaching the Earth's surface, so that at ground level only a fine harmless drizzle of radiation remains.

In addition to familiar electrons, protons, and atomic nuclei, these 'cosmic rays' have been found to contain exotic stuff previously unknown on earth. That is how the positron, the positively charged antiparticle of the electron, was revealed.

In a nutshell the story is that as early as 1923 the very first photographs of cosmic rays showed images left by positrons, but at the time no one realized the fact. Then, after Paul Dirac had predicted in 1928 that such a positively charged version of the electron exists, within four years it was found in the cosmic radiation. First reactions were that the positron was some sort of extraterrestrial, until scientists discovered that it is also produced all the while here on earth, as part of the debris in some radioactive processes. The only reason that no one had noticed was because for positrons ours is an alien world, which quickly destroys them.

The Discovery of the Positron

Positrons had been seen, but not recognized, five years before Dirac's theory appeared. In 1923 Dmitry Skobeltzyn in Leningrad was investigating gamma rays; to make them visible he was using a cloud chamber.

We are all familiar with the vapour trails of jet aircraft, which can remain in the sky for several minutes, providing a record of the plane's movement. The trails consist of small water droplets condensed on the exhaust fumes, creating a long thin cloud. Similar principles apply in the cloud chamber, the first device to produce visual images of particle tracks. It is in effect a glass box containing moist air at low pressure, together with a piston, which can suddenly allow air to rush into the chamber. Water vapour in the air then condenses on any charged particles, revealing their presence and motion by miniature vapour trails. The cloud chamber was to the atomic physicists of the early 20th century like the telescope for astronomers, making visible things that lay beyond normal vision.

Gamma rays do not leave trails directly; like H. G. Wells' invisible man they give themselves away by jostling the crowd. That is how Skobeltzyn planned to catch them. The invisible gamma rays would knock electrons out of atoms in the cloud chamber, whose trails he could see, and from them Skobeltzyn was hoping to infer evidence of the gamma rays.

It worked, but too well. The gamma rays were so powerful that in addition to knocking electrons out of the gas, they were ejecting them from the walls of the chamber as well, which interfered with the measurements that he was trying to make. He then came up with the clever idea of sweeping away the unwanted electrons by putting the chamber between the poles of a large magnet. This thinned the clouds, and the clearer view revealed something utterly unexpected: the magnetic forces seemed to make some of the 'electrons' curve 'the wrong way'.

Today we know that he was seeing positrons, the positively charged 'anti'-version of electrons, but none of this was anticipated in 1923. The anomalous trails were puzzling, but were also a distraction from what he was trying to do. Nonetheless they bothered him.

News about these images spread among the scientific fraternity and five years later Skobeltzyn decided to show them at an international conference in Cambridge. Everyone was as surprised as he had been, but no one could offer an explanation. It was ironic that he was displaying these images in 1928, in Cambridge, the same year and the same place that Dirac would come up with his theoretical prediction of positrons, whose trails would indeed appear to be like electrons 'going the wrong way'. However, as no one at that time had any reason to expect that positrons existed, let alone that they would turn up in Skobeltzyn's experiment, he missed the big prize.*

---

* I have been unable to discover if Dirac attended the conference. However, as he was a mathematician, and it would only be later that his work would turn out to have implications for cosmic rays, it is likely that he was unaware of these developments. Furthermore, Skobeltzyn's presentation may have only made a lasting impact as a result of what was to transpire later. See D. Wilson (1983), p. 562.

A magnetic field deflects the paths of electrically charged particles. The amount of curvature is greater for light or slow-moving particles than for heavier or fast ones, and the direction shows whether the charge is negative or positive, negatives going one way, to the left say, and positives to the right. However, some of the trails went straight through Skobeltzyn's chamber, in what appeared to be straight lines. These were caused by electrons travelling so fast that the magnetic field hardly deflected them before they were gone, indeed much faster than could have come from any source of radioactivity or gamma rays known at the time. In fact they had been knocked out of atoms by the cosmic rays. Although he did not realize it then, he had became the first person to observe the tracks of the cosmic rays themselves. It is almost certain that these trails contained not just electrons but also positrons, but as they did not curve enough for him to know, and as he did not follow this up, he missed the prize for a second time. It was left to Carl Anderson in the USA to make the seminal discovery of the positron in 1932, four years after Dirac's theory had predicted its existence.

Robert Millikan at Caltech, who had won the Nobel Prize in 1923 for his measurement of the electron's charge, had coined the name 'cosmic rays' and had his own theories about the origins of this extraterrestrial radiation. He thought that cosmic rays were gamma rays, the 'birth pangs of creation' as he put it, though quite what he meant is not clear, and that the trails in Skobeltzyn's chamber could be the proof. To tease out what the rays contained you would first need to make them curve, revealing their charge and energy, and to do this required a much more powerful magnet. With strong enough magnetic fields, even the very fastest particles could be deflected. In 1930 Millikan suggested to his research student, Carl Anderson, that he build a magnet powerful enough to deflect the cosmic rays.

This he did with the help of engineers at the nearby aeronautical laboratory. The magnetic fields were ten times stronger than Skobeltzyn had used, and with this more powerful set-up Anderson succeeded in bending the particles' flightpaths. To his surprise he discovered that the cosmic rays contained both negative and positive charged particles in about equal numbers.

As Millikan believed the cosmic rays to consist of gamma rays, which leave no trails themselves, he assumed that the charged particles must have been kicked out of atoms by the gamma rays. His interpretation was that the negatives were electrons and the positives, protons. However the images in Anderson's photographs did not really fit with this. Lightweight particles such as electrons leave thin wispy trails, quite different from the dense trails of bulky protons. All of the trails in Anderson's pictures looked like electrons, and so he suggested that those that curved 'the wrong way' were not due to positively charged particles rushing downwards but were instead electrons moving upwards. Millikan did not like this, and with his judgement skewed by his prejudices on the nature of cosmic rays, insisted that even though the trails were thin and not thick, they must nonetheless be caused by downward moving protons.

Anderson settled the debate by putting a lead plate across the middle of the chamber. If a particle passed through the plate, it would lose energy and so would have a tighter curve afterwards than it had had before entering. This way there would be no argument about whether they were travelling downwards or upwards; it would also determine once and for all the sign of their charges: positive down-mover or negative up-mover.

This indeed answered the question and showed that *both* Anderson and Millikan had been wrong! The trails were due neither to positively charged protons, nor to electrons that were travelling upwards, but were actually trails of downward moving 'positive electrons'. Anderson at least was satisfied, though he still had difficulty in convincing his mentor Millikan, as we shall see later.

It is ironic that Anderson's first sighting of a positron was one that really was moving upwards. It turned out to be a stray produced by a cosmic ray hitting an atom in the air below the lead plate and then bouncing up and through it. This completely confused him, but then he found his first beautiful example of a positive particle that was clearly much lighter than a proton and moving down through the lead plate. He soon found several examples of such 'positive electrons' coming down from above, and had enough confidence to go public. The editor of *Science News Letter* published a photograph of one of the tracks in the December 1931 edition, and coined the name 'positron'. It has been known as such ever since.

In 1931 the received wisdom was that matter is made of atoms and that the atomic menu was simple—electrons and protons. Positrons had no place in this, so where did they come from and what are they? Anderson and Millikan were based on the west coast of the USA and without the instant communications and discussion groups that are the norm today had only a passing awareness of Dirac's work or its implications. Whereas Anderson was the first to identify a positron, it was Patrick Blackett and Giuseppe Occhialini in the Cavendish Laboratory at Cambridge who confirmed its existence without a doubt, and explained where it had come from.

Figure 6. Creation of an electron and positron. A high energy cosmic ray has knocked an electron out of an atom – this gives the gently curving trail from the top to bottom left of the image. There is enough energy also to create an electron and positron, which give the tightly curving counter-rotating spirals at the top of the image. Lower down the picture a further electron and positron are created which depart leaving the inverted vee shape.

Blackett and Creation

Positrons do not exist inside atoms, at least not the atoms of matter that we know on earth, so where had the positrons in the cosmic rays originated? Anderson didn't know and it was Blackett and Occhialini that same year who found the answer: positrons are not extraterrestrial invaders but are created in the atmosphere by the cosmic radiation itself.

Blackett had been working with a cloud chamber in Rutherford's group at Cambridge's Cavendish Laboratory. He had a passion for gadgets and devised a chamber that was ready for action every ten seconds or so, and took photos on ordinary cinematograph film. Between 1921 and 1924 he accumulated more than 20,000 pictures of trails made by alpha particles—a product of radioactive nuclear decays—which were bombarding nitrogen gas in the chamber. Occasionally an alpha particle would collide with the nucleus of a nitrogen atom, merging with it in such a way that the nitrogen was modified into the seeds of another element. By capturing nuclear transmutation on film in this way, Blackett had begun to make a name for himself.

It was in 1931 that Giuseppe Occhialini arrived at the Cavendish. His speciality was detecting nuclear radiation using Geiger counters. He and Blackett compared notes and began to realize that by combining their expertise they could turn the cloud chamber, which until then had been something of a hit and miss affair, into an efficient instrument.

The idea was brilliantly simple. Blackett's cloud chamber worked automatically taking pictures over and over again, waiting for the lucky chance when something happened: most of the pictures showed nothing interesting, only about one in every twenty contained any tracks. The Geiger counter's strengths and weaknesses complement those of the cloud chamber; a Geiger counter fires when a charged particle passes through it, but reveals little if anything about what actually triggered it. Their big idea was to put one Geiger counter above a cloud chamber, and another one below. If both of them fired simultaneously, it would be very likely that a cosmic ray had passed through the chamber. By connecting the Geiger counters to a relay mechanism, the electrical impulse from their simultaneous discharges triggered the cloud chamber and a flash of light captured the tracks of the cosmic rays on film. The key feature was that the trails live on after the ray has passed—by the time the photo was taken, the ray was long gone, but the all-important drops were still there in the gas.

Whereas Blackett's success rate previously had been only one in twenty, now it leaped to four out of five! He and Occhialini took their first photographs by this method during June 1932 and then accumulated nearly a thousand pictures in the late autumn of that year. They noticed that a few of the tracks that appeared at first sight to be electrons, were actually curved the wrong way in the magnetic field. Blackett talked to Dirac about them.

Everything was in place for the Eureka moment where Dirac would dramatically exclaim that this was a positron, the proof of his theory. But no. Somehow Blackett and Dirac failed to put two and two together. Dirac's penchant for cautious logical conversation may have been the reason; he was hardly a person who would adopt the modern 'hard-sell' of an idea. Or perhaps Blackett did not appreciate the depth of Dirac's theory, or simply did not take it seriously. Whatever the reason, Blackett and Dirac parted with neither of them aware of the precious truth that had been before their eyes. As Skobeltzyn had missed the great prize, so now the discovery passed them by too. It was only when they heard of Anderson's discovery that Blackett and Occhialini finally realized what they had.

But luckily for them they had more, something that neither Skobeltzyn nor Anderson's hit and miss experiments had chanced upon. Many of their pictures showed up to twenty particle tracks diverging from some point in a copper plate just above the chamber like water from a shower. The powerful magnetic field throughout the chamber curved the tracks, showing that roughly half of the particles were negatively charged and the rest positively charged. Blackett and Occhialini realized that as positrons do not occur naturally on earth, the appearance of equal numbers of positrons and electrons must be because they were being produced by some invisible high energy cosmic radiation. The message was that the positrons were being formed as a result of collisions between the cosmic rays and atoms in the chamber.

The cloud chamber had glass sides encased in copper and the showers were the result of cosmic rays hitting the metal. A single electron in the cosmic rays was enough to make a cascade of electrons and positrons this way. The intense electric fields within the copper atoms made the passing electrons radiate gamma rays, and provided these gamma rays had enough energy, they in turn produced pairs of electrons and positrons. Albert Einstein's equation, $E = mc^2$, implies that energy (E) can be converted into mass (m)—radiation into matter—and Blackett and Occhialini had for the first time demonstrated the creation of matter, and antimatter, from radiation; they had proved that Anderson's new particle was not some weird extraterrestrial interloper.

The final irony in this drama is that they almost beat Anderson to the full credit. Anderson had spent time trying to convince his supervisor, Millikan, that he had indeed found a positive version of an electron and was not simply seeing protons. Blackett and Occhialini's work proved the issue beyond any doubt, and reluctantly even Millikan had to agree that Anderson had been correct. Blackett and Occhialini's paper was sent to the *Proceedings of The Royal Society* in February 1933. Fortunately for Anderson, he had been confident enough to go public with his tentative result in 1932, Millikan's scepticism notwithstanding, following up on the picture in *Science News Letter* in the preceding December. Fortune favours the brave.

Positrons on Earth

Once Dirac had pointed the way and news of the discoveries by Anderson, Blackett, and Occhialini spread rapidly around the world of science, positrons turned up all over the place. Physicists immediately looked through their old photographs from cloud chambers and found evidence for positrons, which they had previously overlooked. Many people, had they been brave enough, might have had their names instead of Anderson's immortalized in the annals of science. Among those who had missed the positron were Irene and Frederic Joliot-Curie. Irene, daughter of Marie Curie, and her husband Frederic Joliot had already missed one Nobel Prize by having been the first to produce neutrons in January 1932,[1] and having mistaken them for gamma rays. Now they realized that they had missed the positron also: Anderson was awarded the Nobel Prize in 1936 for the discovery. However, their luck was about to turn as they won the Nobel Prize for chemistry in 1935 for the production of short-lived radioactive nuclei; one of the applications of their work is the production of nuclei that spontaneously emit positrons.

When Henri Becquerel stumbled upon radioactivity in 1896, he had discovered that the nuclei of uranium atoms can spontaneously change; today we know that this is due to a neutron in the nucleus turning into a proton, the electric charge overall being balanced by emission of a negatively charged electron. Following the discovery of the positron, it was natural to suppose that there might be nuclear decays where a proton turned into a neutron, the electric charge being taken away by a positron.

The idea grew that radioactivity could produce positrons as easily as electrons. The major practical difference between the two lies in what happens next. An ejected electron may flow as electric current or join in the dance of planetary electrons in neighbouring atoms, later to initiate chemical reactions and countless other adventures in the future of the universe. A positron by contrast is a stranger in our land and isn't long for this world. It finds itself surrounded by matter containing hordes of negatively charged electrons. Momentarily one of these electrons partners the positron in a cosmic dance of doom as they encircle one another and, within a microsecond, mutually annihilate in a flash of light. It is this that in recent years has become the key to the practical use of positrons.

Positron emission is natural and common; it is the ability of certain nuclear varieties to emit them that has become so useful in medicine and technology. Some examples of such nuclei are carbon-11, nitrogen-13, and oxygen-15, which are radioactive forms of common elements in the body and can be used, along with positron emission, to trace bodily functions such as those in the brain. The basic principle is that when the nucleus emits a positron and the latter annihilates with a nearby electron, two gamma rays can emerge almost back to back. This pair can be detected using electronic circuitry developed in particle physics, enabling you to locate the emitting nucleus very accurately. Now for the applications.

When you are thinking, various parts of your brain are active to different degrees. Being active uses energy that is supplied to the brain as chemical sugars in the bloodstream. If we could measure the concentration of sugar within the brain it would give some indication of the brain's activity. Chemists can incorporate radioactive atoms into sugar molecules and those sugars can be ingested and distributed within the body to the regions that are active, such as the heart, lungs, muscles, and the brain. The essential idea, which has proved so useful in medical diagnostics, is to use sugars that emit positrons. The positrons are immediately annihilated by the ubiquitous electrons in nearby atoms. We can tell where in space the annihilation took place, and hence where the sugar was located, simply by using special cameras to detect the gamma rays that come flying out.

By surrounding a patient's head with a halo of cameras, images of the brain can be built up in slices. This technique is known as positron emission tomography, or PET. The particular isotopes of interest tend to be rather short-lived* and therefore have to be prepared near to the patient. They can be made by bombarding suitable elements with protons from a small accelerator.

So today, Dirac's arcane prediction of antimatter is being used to save lives. Positron annihilation is also useful for studying materials. One example involves annihilation in metals that can reveal the onset of metal fatigue much sooner than other techniques can. This has been used in testing aircraft turbine

---

* For example oxygen-15, used to investigate oxygen metabolism, has a half-life of only two minutes.

blades, enabling safety margins to be narrowed and profits increased.

Scientists are studying the chemical properties of antimatter by binding positrons to ordinary atoms. As an electron and proton form atoms of hydrogen so an electron and positron can form an atom of 'positronium', which lives for less than a millionth of a second before self-annihilation. Molecules of positronium have even been formed and there is speculation that dense collections of these molecules might form the basis of a gamma-ray laser.[2]

So antiparticles, in the form of positrons, are familiar and put to use daily. They are less familiar than electrons merely because they are so outnumbered and hence rapidly killed off. And as we

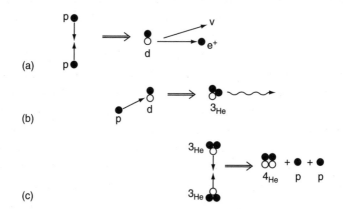

Figure 7. Positrons and energy production in the sun. Two protons denoted by p fuse to make a deuteron (consisting of a neutron, the white circle, and a proton, the dark circle) together with a positron and a neutrino. In (b) another proton hits the deuteron, converting it to helium-3 and a photon. In (c) we see the consequence of two of these processes: two nuclei of helium-3 combine to make one of helium-4 and two protons.

already said on page 13, daylight is the end product of positrons, the seeds of antimatter, that were produced in the solar furnace a hundred thousand years ago, and the illuminations for the far future are being prepared by positrons that are being made right now.* (See figure 7.)

---

* Solar fusion effectively turns four protons into a nucleus of helium, plus two positrons and neutrinos. The positrons annihilate to produce photons. The mass of helium is less than the sum of the four proton masses and the excess mc^2 is energy that also emerges at the surface as visible light. About 10% of visible light originated in positron annihilation.

ANNIHILATION

Neither Matter nor Antimatter

There's matter, like the electron; antimatter, like the positron; and then there are things that are neither matter nor antimatter. The most familiar example of something that is beyond substance is electromagnetic radiation. All electromagnetic radiation, from gamma rays through X-rays and ultra-violet to visible light, infra-red, and radio waves, consists of photons of different energies. Matter and antimatter can cancel one another out, their annihilation leaving non-substance in the form of photons; if the conditions are right this sequence can happen in reverse where photons turn into pieces of matter and antimatter.

Pure energy, that concept beloved of scientists when auditing the accounts of natural processes, also is non-substance; it can change from one form to another, such as electrical, chemical, or motion, and it can transubstantiate into matter and antimatter.

Einstein tells us how much substance can condense from energy; that is what $E = mc^2$ is all about. The minimum amount of energy to make an electron and a positron will be $2mc^2$: one lot of mc^2 is enough to make a stationary electron, and one lot to make a positron. Having been created stationary, they will almost certainly immediately annihilate one another, freeing the energy that was momentarily trapped within them. To give the positron some chance of surviving you need to have more energy than this minimum; the 'excess' becomes kinetic energy, motion, so that at their birth the electron and positron will rush apart and escape.

The photon of light is just one example of over a hundred known examples of particles that are non-substance. Such entities are known as 'bosons', after the Indian physicist Satyendranath Bose. By contrast, substantial particles that are basic pieces of matter or of antimatter are known as 'fermions', after the Italian Enrico Fermi. The behaviour of fermions is described by Dirac's equation; bosons follow different rules. In a sense, Dirac had been lucky. His aim had been to shape an equation for particles that have mass, and to confront the problem of positive and negative energies. In 1928 the only massive particles known were the electron and proton, each of which by chance is a fermion; the only other particle to have been identified, the photon, was a boson, but without mass. Twenty years after Dirac had revolutionized science with his equation, a massive boson, the 'pion', was discovered in cosmic rays. Had the pion been known in 1928, we can only imagine whether Dirac would have pursued his equation so intensely, if at all.*

---

* The so-called 'alpha particle' emitted in some radioactive decays is also a boson. However, this was known to be the nucleus of a helium atom and hence not as fundamental as these other particles.

The universe is built from basic particles, trapped in a never ending terpsichore by the natural forces, of which gravity and the electric and magnetic forces ('electromagnetic force') are the most familiar. These act over large distances, effectively infinite when compared to the scale of atoms. Gravity entraps the planets in their orbits around the sun, while the swirling electric currents within the earth's molten core give rise to magnetic fields that can swing a small compass needle, guiding lost travellers towards home. At least that is what used to be the case; today they would more likely rely on a GPS system, but the underlying principle is similar: communication with the satellite is by means of radio waves, electromagnetic radiation, a manifestation of these same ubiquitous forces.

As you watch a magnet attracting pieces of metal, or the compass swinging around towards the north pole, you might wonder what agent communicates between them. We can give it a name, 'the electromagnetic field', but this is not really an explanation; we are just inventing a label for the weird phenomenon of actions over remote distances. One of the results of Dirac's work was the discovery that the electromagnetic field itself is ruled by quantum theory. Photons are particle-like bundles of electromagnetic radiation, and they transmit the electromagnetic force as they flit between one charged particle and another. An electron oscillating back and forth in a radio antenna in London can cause a similar response within your radio at home, the communicant being electromagnetic waves—radio waves that are also the movement of photons. Motion in one location has given rise to motion elsewhere; photons travelled across the intervening space; the force is with you.

In modern 'quantum field theory', not just the electromagnetic force but all the forces are transmitted by bosons. What the photon does for the electromagnetic force, so the 'graviton' is believed to do for gravity. No one has yet detected a graviton, but few doubt that it exists and that some day it will be found. There are two other forces, which are also known to be transmitted by bosons. These forces are less familiar, because they act principally in and around the atomic nucleus and are only revealed by tools sensitive enough to resolve such fine scales. They are known as the strong force and the weak force, their names summarizing their apparent strengths relative to the familiar electromagnetic force.

The strong force constructs protons and neutrons out of more basic pieces known as quarks (of which more later) and glues the atomic nucleus together. The weak force makes the sun shine, and is crucial in building the elements, without which the earth, and we, would not exist. It is this force that slowly eats away inside atomic nuclei, eventually transforming their constituents into more stable combinations. Thus in the sun, where protons are the fuel, the weak force gradually converts four of these protons into a compact cluster that is the nucleus of helium, which consists of two protons and two neutrons. In doing so, the weak force has transmuted two of the protons into neutrons, their positive electric charge being taken away by—positrons. In five billion years about half of the sun's fuel, its protons, has been changed in this way. This gives an idea of how weak the force is when acting within the solar furnace, for which we may be thankful: the sun has lasted long enough for intelligent life to have emerged, but has nonetheless burned fast enough to provide the conditions for life to have developed at all.

The strong and weak forces have fascinated physicists for over half a century since their existence was first recognized. Today we understand how they work, and in chapter 6 we shall see how it was the use of antimatter that revealed their secrets. They too are transmitted by bosons. 'Gluons' are what glue quarks to one another to build up protons and neutrons and then 'pions' help to grip the latter to build up the nuclei of atoms. The weak force manifests itself in two distinct ways, and there are bosons that transmit each of them. One is similar to the electromagnetic force but much more feeble in strength, and is transmitted by an electrically neutral boson known as the Z^0 (the superscript zero denoting its lack of charge). This Z^0 is like a photon except that it is very massive, more so even than an atom of iron; it has been whimsically referred to as 'heavy light'. The second way that the weak force can act is in swapping around the amounts of electrical charge of the participating particles. For example, in transforming a proton into a neutron within the sun, the weak force has taken the electric charge from the proton and passed it on to a positron. And where did that positron come from? It was created from the energy carried away by the carrier of the weak force, known as W^+. In this case the superscript denotes that the W had positive electric charge. The W can also exist negatively charged, as when a neutron decays. In this case the neutron's zero charge has turned into a positive (the proton) and a negative (the W^-), the negative electric charge of the W^- being passed on to an electron.

All of these agents that transmit forces are non-substance, neither matter nor antimatter; they are all 'bosons'. They act on particles of matter or antimatter, and can themselves turn into counterbalancing pieces of these two forms of substance, which are all 'fermions'. So nature seems to have provided two varieties of particles: force carriers, which are bosons, and the basic bricks of substance, which are fermions. Bosons can come and go; fermions eventually decay down to their most stable forms, electrons and combinations of protons and neutrons, at which point they are only at risk from their antimatter doppelgangers.

The battle between matter and antimatter in the universe was fought fourteen billion years ago, and matter won. Fermions give rise to structure, they have stability and lead to life. We are formed from atoms that have existed for billions of years, it is only now that they are configured in combinations that think they are us. We breathe in oxygen, exhale carbon dioxide, grow and die, but our atoms will go on. Their basic pieces will recombine with infinite variety into the distant future, so long as they do not meet antimatter.

More Antiparticles

Dirac originally wrote his equation in order to explain the electron. However it is the script that all fermions read and applies equally well to a proton or a neutron. As the equation implied that the electron has a negative energy version, which Dirac successfully interpreted as a positively charged positron with positive energy, so it implied that the proton and neutron also have antimatter counterparts, the antiproton and antineutron. The antiproton has the same mass as a proton but negative instead of positive charge. The antineutron likewise has the same mass as a neutron, and zero charge. Since the charge of the antineutron is the same as that for a neutron, what property distinguishes between them?

Although the neutron has no electrical charge overall, it contains charge within. As we shall see, both the proton and neutron have a small but measurable extent within which are swirling electric charges, positives and negatives, which add up to the total that we call the charge of the proton or neutron. Although a neutron's total charges add to zero, their motions produce

69

swirling electric currents and magnetism, which can be sensed by watching how a neutron moves in a magnetic field. Inside an antineutron, these individual electric charges are each reversed in sign, its inner currents cavorting like a mirror image of those in a neutron. As a result, the magnetism, its north and south poles if you will, are reversed. In a magnetic field, the paths of neutrons and antineutrons are like mirror images. As we shall see later, the points of electric charge that combine to form protons and neutrons are themselves little particles, known as quarks. A proton or neutron is made of quarks; their anti-selves are made of antiquarks.

After the positron had been discovered, the challenge was to verify if the other piece of an atom of antihydrogen, the antiproton, also exists. The problem was that as the proton is nearly two thousand times heavier than an electron, so would an antiproton be that much heavier than a positron, which means that much more energy is needed to make it. While antiprotons do occur in cosmic rays, they are much rarer and harder to identify.

By 1950, cosmic rays had revealed further novel particles. These included the muon, which is a heavy version of the electron; the pion, which is a boson that weighs about one-seventh as much as a proton; and particles such as the kaon which were dubbed 'strange' particles because they were unexpectedly long-lived; but not the antiproton. It was received wisdom that the antiproton exists, so confident were the physicists in the truth of Dirac's insight. An ambitious plan took hold at Berkeley in California to build an accelerator that would speed protons such that when they smashed into a target, there would be enough energy to produce an antiproton. The challenge was to make the machine and to design a means of detecting and identifying the antiproton without any doubt.

The machine was known as the BeVatron—'BeV' being the shorthand for a 'billion electron volts', the amount of energy locked into an antiproton. When energy turns into massive particles they emerge in pairs, a particle being matched with its antiparticle, so the BeVatron was built with enough power to produce an antiproton in conjunction with a proton, and there was confidence that it would do so. However, all conjectures need to be proved experimentally and this was tricky because antiprotons would be very rare and overwhelmed by the production of lighter particles such as showers of electrons and positrons, and of pions.

Several ideas on how to isolate the antiproton 'needle' from the particle 'haystack' were presented to the committee of scientists who would decide which proposal looked the most convincing. A small team consisting of Owen Chamberlain, Emilio Segre, Clyde Wiegand, and Tom Ypsilantis won the competition and had first turn on the new BeVatron. Their idea worked, the experiment was a success and in 1955 they announced the discovery. One of the other teams led by Oreste Piccione that had entered the competition also gained success themselves with the discovery of the antineutron in 1957. So thirty years after Dirac had first produced his seminal prediction, the basic pieces of the antiworld were all in place: positron, antiproton, and antineutron.

That was the end of the beginning in the story of antimatter, but it was only the start of years of discord, which led to litigation. In 1959 the Nobel Prize for physics was shared by Chamberlain and Segre, who had led the experiment, for their role in the discovery of the antiproton. The real fracas concerned Oreste Piccione, who had been in the team that had discovered the antineutron. It was not so much that he felt the antineutron deserved a prize also, since in everyone's mind finding the antineutron was merely icing on the cake, as that he felt he should have had a share of the antiproton prize. As he was not one of the quartet, you might wonder why.

When the BeVatron committee began their deliberations on who would have the first opportunity at the new machine, Piccioni had come up with some clever ideas on how to capture the elusive antiproton and had included them in his proposal. However, on balance the proposal of Chamberlain and Segre's team seemed the better bet, and so they went first, Piccione second. If that had been the end of the story it simply would have been a case of bad luck. However, at least in Piccione's opinion, his ideas were incorporated in the rival team's experiment and proved instrumental in them hitting the jackpot. This continued to rankle him to the extent that in 1972 he instigated a lawsuit claiming damages for being wrongfully excluded from the Nobel Prize that had been awarded thirteen years earlier. It proved to be unlucky thirteen: it was deemed that too long had passed between the legal action and the event in question, and the case was duly thrown out on a statute of limitations.

That was the legal outcome; whether he would have been successful had he not delayed, we shall never know. In the scientific community opinions were divided. Some felt Piccioni deserved a share; at the other extreme some said that none did, or perhaps the committee that had allotted time for the experiments, or those who designed the machine with the antiproton in mind should have won the prize as it was their 'insight' that recognized its singular importance. For the scientific community at least, antiparticles no longer appeared to hold any special interest. However, this was about to change with the discovery of a deeper layer of matter, 'quarks', and of antimatter, 'antiquarks', which would eventually lead to an explanation of how matter had emerged from the Big Bang.

Quarks—and Antiquarks

When Dirac came up with the idea of antimatter he knew only of the electron and proton. Even after the discovery of the neutron, remarkably in the same year as Dirac's positron was found, the particle menu was still relatively straightforward. However, within thirty years so many particles had been discovered in cosmic rays and in the new particle accelerators, that had Dirac produced his prediction at that time it might have caused little stir: one more particle; so what?

The accelerator at Berkeley, designed to produce the antiproton, also added to the proliferating menu of particles. All these new particles were unstable, some living for no longer than it would take a light beam to cross an atomic nucleus. In effect this is like saying that the particle died the moment it was created, as Einstein's theory of relativity implies that information cannot travel faster than the speed of light, and it would take that brief instant for the entire particle to have formed and fragmented. Other particles were discovered that lived for longer, though even by this we mean less than a billionth of a second, or about the amount of time it would take light to cross your hand. You might wonder how one could know of something so ephemeral. The answer is the power of modern electronics, and the fact that when these particles travel at near the speed of light, they can traverse measurable distances during their brief lives.

Any particle that has electric charge will knock electrons from (or 'ionize') atoms in the air as it bumps into them. If the air is damp, a vapour trail will form as the particle passes. The cloud chamber revolutionized understanding of atomic particles, including the discovery of the positron, in the first half of the 20th century; the invention of more powerful tools made it a museum piece in the latter half.

In 1952 at the University of Michigan, Donald Glaser was musing about the way bubbles form in a glass of beer. This inspired him to invent the bubble chamber, which quickly became a remarkable means of revealing the dance of subatomic particles. Whereas in the cloud chamber particles formed bubbles of liquid in a surrounding gas, in the bubble chamber they formed bubbles of gas within a liquid. The images of bubble trails, spiralling in magnetic fields, splitting as one particle decays, spawning progeny like a parent passing on their genes, form beautiful pieces of artwork as well as revealing profound truths to those that have learned how to decipher them.

It was with the bubble chamber revolution that whole families of new particles showed up. Nearly ten years elapsed before order began to emerge.

The proton and neutron were soon joined by particles that in many ways appeared to be heavier versions but with properties that led to them becoming known as the 'strange particles'. Some were more strange than others. Then there were particles that did not have the special character known as 'strange' even though they were peculiar and tantalizing. Their names covered the Greek Lambda (Λ), Omega (Ω), Sigma (Σ), Xi (Ξ), and Delta (Δ), and as these began to be used up so the naming continued into lower case phi (ϕ), eta (η), omega (ω), and rho (ρ), eventually spilling over into the A, B alphabet. The famous statement of Rutherford that science consists of physics, all else being stamp collecting, was turning into extreme irony.

As more and more particles showed up, common features grad-ually began to be discerned among some of them, which hinted that they were not all independent but instead belonged to various families. This was reminiscent of what had happened with the atomic elements in the previous century. Mendeleev had noticed regularities among the elements, which he encoded in his Periodic Table. Later the explanation for this periodicity was found: atoms are made of a few common constituents, electrons encircling a nucleus of protons and neutrons. The proof that atoms are really like this, made of smaller pieces, came first from J. J. Thomson liberating electrons from within them, and then from Ernest Rutherford who discovered the existence of the atomic nucleus. The case with protons and the multitude of short lived particles was similar in spirit to the atomic case, though different in details.

Rutherford had realized that there was some hard centre to the atom from the fact that when alpha particles hit atoms, occasionally the alphas recoiled violently as if they had struck some compact object within: the atomic nucleus. A similar set of events took place years later, on a grander scale, with the discovery that the proton, neutron, and their many relatives were not the basic seeds of matter, but were made of smaller particles called quarks.

Whereas Rutherford and his assistants had been able to discover the atomic nucleus in an experiment on a table top in a room at Manchester University, to invade its protons and neutrons required an accelerator that is 3 kilometres long. When beams of electrons emerged from the accelerator, at Stanford south of San Francisco, they hit a target of hydrogen and ploughed deep within the protons that are at the heart of each atom. Occasionally the electrons bounced sharply from their path, much more violently than would have been the case if the protons were just a miniature ball of electric charge. As had been the case for the nuclear atom, so it was for the proton: the electric charge of a proton is not smeared smoothly throughout its volume but

75

is instead concentrated on three much smaller particles within, known as quarks. Indeed, what we call a proton is nothing more than these quarks rushing around, trapped within a permanent prison extending no more than one millionth of a billionth of a metre in size. Like an ant hill that at first sight appears to be a well defined brown lump but on closer inspection is seen to be a seething mass of small creatures, so does the proton from afar appear to be a compact ball of charge but close up is revealed as a jumble of quarks.

Quarks glue to one another strongly in threesomes, making protons, neutrons and many other members of the particle alphabet soup. The two varieties of nuclear particle, proton and neutron, are themselves made of two varieties of quark, known as the 'up' and 'down'. These seeds of matter have electrical charges in amounts that are fractions of a proton's charge. Where the proton has one unit of positive charge, an up quark has $2/3$ positive and a down has $1/3$ negative. Two ups and one down then make a proton $(2/3 + 2/3 - 1/3 = 1)$; two downs and one up add to zero charge $(2/3 - 1/3 - 1/3 = 0)$, which is a neutron.

There are combinations of all three up quarks or down quarks, which are short lived particles known as Delta. Add to this pair a third type of quark known as 'strange', which has the same electrical charge as a down quark $(-1/3)$ and is effectively identical in all respects except that it is about 20 per cent heavier, and you have everything needed to explain the strange particles; the more strange quarks there are in each threesome, the stranger is the particle that is made. The proton and neutron contain no strange quarks; the heavier and slightly strange particles known as the Lambda (Λ) and Sigma (Σ) contain one; doubly strange particles, known as Xi (Ξ) contain two, and the strangest of all is the Omega (Ω) which consists of three strange quarks.

Dirac's equation applies to quarks just as it does for electrons and protons, and with the same implications for antimatter. As a positron is the antiparticle mirror to an electron, so is an antiquark to a quark, having the same mass, the same size, and the same amount of electric charge as the quark, except that the sign of the electric charge is opposite. So the anti-up (take your choice as to whether you prefer to call it an up-antiquark, or an anti-up quark, or an anti-up anti-quark; there is no agreed convention) has charge $-\,^2/_3$ instead of $+\,^2/_3$; the anti-down has charge $+\,^1/_3$ instead of $-\,^1/_3$. As two ups and a down make the positive proton, so two anti-ups and an anti-down make the negative anti-proton. Likewise two anti-downs and one anti-up can clump to make an antineutron. The strange particles with Greek names, such as Lambda, Sigma, Xi, and Omega each have an 'anti' counterpart; replace each variety of quark by the corresponding antiquark and you have the anti-Lambda, anti-Xi and so on. All of these have been produced in experiments, where the spare energy in a collision between a beam of protons emerging from an accelerator and a target in the laboratory has produced new particles and antiparticles. To the best accuracy that we can achieve, each and every one of these appears to be a perfect mirror counterpart of its particle analogue, the yin to its yang.

When Quark Meets Antiquark

The strong force that grips quarks or antiquarks in trios can also attract a single quark to an antiquark. Many of the particles found in cosmic rays and at accelerators, such as the pion and the 'strange' kaon, are tiny bundles of a quark and an antiquark. Such a combination is neither matter nor antimatter; it contains samples of each, a quark of matter and an antiquark of antimatter. When a quark and an antiquark are confined to a miniature universe whose extent is only a thousandth part of a billionth of a metre, they will meet and be annihilated almost immediately. Thus the pion and kaon do not survive for more than a brief moment.

Experiments have shown what happens when a proton meets an antiproton. Sometimes they just drift into one another, other times they slam into one another at speed and the end-products vary. The faster they collide, the greater their energies, the more pions or gamma rays are produced in the explosion. Scientists have learned so much from these experiments that if someone successfully makes an antimatter power source, or bomb, or if a lump of antirock from outer space hits the atmosphere, we know enough to predict the likely outcomes.

What these experiments have shown is that annihilation is not instantaneous. Instead, a proton and antiproton do a brief dance of courtship before entering on their final fateful coupling. Imagine protons in a lump of matter as the antiproton approaches. The positive charge of the proton creates an electric field that spreads out into space over atomic dimensions. While such distances of about a ten billionth of a metre appear small to us, they are nonetheless some ten thousand times larger than the size of either the proton or antiproton themselves. If the antiproton approaches the proton relatively slowly, it will be entrapped by the attraction of opposite charges and start to orbit around the proton much as an electron might do in a conventional atom. Initially this orbit will be far away, but the antiproton will lose energy, dropping from outer orbits to inner ones, emitting gamma rays as it does so. It is these gamma rays, which can be detected and their energies measured, that are like the skid marks at the scene of a crash, revealing the sequence of events.

The antiproton tumbles into a close-in orbit, and comes within the range of the strong force, which is irresistible for protons and antiprotons. This dance may have lasted as long as one-hundredth of a second, but once ensnared by the strong force doom is almost instantaneous. The news of the catastrophe travels at the speed of light across the proton and the antiproton, and within less than a billionth of a billionth of a second they are gone, leaving gamma rays and pions. Then these too are gone; the pions, short-lived constructs of a quark and an antiquark, self-destruct, turning into yet more gamma rays, or into electrons, positrons, and ghostly neutrinos, all of which carry away the energy from the annihilation.

The unique opportunity that annihilation offers for releasing energy from matter, is a mixed blessing. In our material world, antimatter is all-destructive. For antimatter to be useful we must contain it, away from any pieces of matter, and for long enough until it is ready for use. How that challenge has been solved is the next part of our story.

STORING ANTIMATTER

The All-Destructive Substance

The solution to my father's question about how to store the all-destructive substance was found by Bruno Touschek. Since anti-matter will destroy any material object, it must be kept in a cage without material walls. The solution is to have a vacuum that is better than in outer space with magnetic and electric fields that confine the antiparticles, positrons, or antiprotons, as circulating beams.

That is in effect what was done at particle physics laboratories such as CERN, where a ring of magnets 27 kilometres in circum-ference guided bunches of positrons around an evacuated tube for weeks on end. Moving within 55 metres per hour of the speed of light these positrons would have survived for as long as the electricity powering the magnets kept them away from the walls of the vacuum tube, or until they collided with stray atoms of gas inside it.

We shall learn about this later. The question that concerns us right now is how antimatter can be stored, transported, and used in reality. Clearly it is not practical to build 27 kilometre rings of magnets, let alone transport them around.

Nor does one need to. CERN's enormous ring was a pinnacle of scientific achievement, designed specifically to make and control beams of antimatter as near to the natural speed limit, 300,000 kilometres per second, as possible. The original idea and technology for this brand of physics had come years earlier, in 1960. The inventor was the Austrian Bruno Touschek, though at the time neither he nor anyone foresaw that here was the birth of an antimatter store.

During the Second World War Touschek had been working on radar in Hamburg. One of his colleagues was a Norwegian, Rolf Wideroe, who twenty years earlier had come up with an idea for accelerating particles by a series of small pushes from relatively low accelerating voltages. In Wideroe's prototype, electric fields accelerated the particles in straight lines. Next, Ernest Lawrence, an American, made use of magnetic fields to steer the path of the particles into a circle such that they passed through the same accelerating gap many times. Lawrence's 'cyclotron' led to the birth of modern high energy physics, and a Nobel Prize. Wideroe's basic idea has remained the principle behind even modern accelerators. It was his next idea that led to the breakthrough.

In 1943 Wideroe applied for a patent on a scheme to store and then collide particles travelling in opposite directions around the same orbit. His application was turned down on the grounds that it was 'obvious', even though it would be another fifteen years before others would put it to use. If you fire two particles at one another they are much more likely to miss one another than to hit. However, if you accumulate lots of them, store them until needed and then fire the two intense beams at one another, there is a reasonable chance that particles in the two opposing beams will find themselves at the same place at the same instant.

The first application of the 'obvious' idea was in 1959. An American team built two 'storage rings', using magnets to steer electrons around them. In one ring the magnetic fields steered the electrons clockwise, and in the other the fields were reversed so as to send the electrons anticlockwise.

It was at this point that Touschek remembered his wartime conversations with Wideroe, and suddenly had his own big idea: positrons have the same mass as electrons but the opposite sign of electrical charge, so a magnetic field that steers electrons to the right, say, will steer positrons to the left. Instead of two sets of magnets to send electrons in opposite directions, why not have just one set of magnets that would send electrons one way and positrons the other! The two beams will follow exactly the same counter-rotating paths provided they have the same energies.

Touschek and a group of colleagues at the Frascati Laboratory near Rome designed and built ADA, for *Anello d'Accumulazione* (accumulation rings). The whole thing was just a metre in diameter. They successfully stored electrons and also positrons, the first time ever that antiparticles had been tamed. Then this machine that could fit in a suitcase was transported to Orsay near Paris, which possessed a more intense electron beam.

It was there in 1963 that intense beams of positrons were successfully stored, and also made to pass head-on through beams of electrons. Occasionally individual electrons and positrons in the beams would collide and, in so doing, the pair would annihilate in a flash. So you had your choice: here was a way to store antiparticles if you wished, or to collide and annihilate them if you preferred.

During the next thirty years physicists built larger and larger storage rings of electrons and positrons whose beams were at ever higher energies. By crashing them into one another and annihilating them, they discovered that this was a wonderful way of learning about the origin and nature of matter, and several breakthroughs led to Nobel Prizes. The largest such machine ever built was that at CERN known as LEP—Large Electron Positron collider—with which we started this chapter. Not all work is headed towards gargantuan machines however. In recent years smaller examples have been built in Stanford, in Japan and even at the origin of it all, Frascati, to make electrons and positrons collide in specially selected conditions where it is hoped that any differences between matter and antimatter might be revealed. We shall learn more of this in chapter 8.

The fact that machines like these work at all shows the remarkable symmetry between matter and antimatter. Beams of electrons and positrons travelled around LEP, over and over again, and arrived at their rendezvous on time, which testifies to their identical response to the guiding magnetic forces. This is a result of their precisely counterbalanced electric charges and their identical masses causing them to follow their pre-assigned paths in opposite directions. Similarly we can compare the times for a proton or an antiproton to complete a circle in the magnetic field, from which we know that the proton and antiproton are alike to better than one part in a billion.

Storing Antiprotons

Touschek had tamed positrons; in Russia Gersh Budker decided to see if he could do the same for protons and, later, antiprotons. Protons and antiprotons are nearly two thousand times more massive than electrons and positrons, and the energy required to make them is correspondingly greater also.

Making antiprotons is not a problem however, so long as you have enough energy available, and this was first done in 1955 as we saw on page 71. But, controlling them once you have made them is a big challenge. First fire a beam of protons into a block of metal. About once in every 250,000 collisions, kinetic energy is converted into mass in the form of a new antiproton–proton pair. These antiprotons are travelling close to the speed of light, rushing all over the place. Magnetic fields that had been able to focus positrons into stable orbits were unable to control the wild antiprotons, which would fly sideways out of their arcs, smash into the walls of the ring and be annihilated.

Some way was needed to tame them; metaphorically they had to be brought into step or, in the jargon, 'cooled'. Budker's idea was to pass the antiprotons through clouds of cold electrons. Although electrons are matter and antiprotons are antimatter, they are no danger to one another: electrons are destroyed by their antiparticle, the positron, while the antiproton is at risk only from protons or neutrons. The antiprotons' erratic wobbles were gradually smoothed out as, in effect, their energy, or 'heat', was transferred to the electrons. By 1974 Budker had succeeded in making and cooling antiprotons, but not in sufficient numbers to make an intense beam.

In 1979 the Nobel committee awarded their physics prize to Sheldon Glashow, Abdus Salam, and Steven Weinberg for their theory that united electromagnetic and weak forces into a single 'electroweak' force, even though its most dramatic prediction, namely the existence of W and Z particles, had not been proved. The plan at CERN, driven by the experimental genius and energy of Carlo Rubbia, was to produce them, and to do so would require a whole new technique—high energy annihilations between protons and antiprotons. Theorists had calculated that in such circumstances it would be possible to produce not just electromagnetic radiation, light, but also the quantum bundles known as W and Z, which are the agents that transmit the weak forces of radioactivity.

Rubbia planned to do this by rotating beams of protons and antiprotons in opposite directions around CERN's 'Super Proton Sychrotron', known as the SPS. That was the end game. To start it all off, beams of protons circulated in CERN's older and more modest, less 'super', machine known as the PS. The next step was to extract the protons, speed them up to make higher energy beams, and then inject them into the SPS. That was the easy part. The bigger challenge was to tame antiprotons and get them successfully into the SPS.

The conundrum was solved by the Dutch engineer, Simon van der Meer, which led him to a share of the Nobel Prize in 1984. His insight was that when particles go around a curve, it takes longer for them to traverse a semicircle than to send a signal at the speed of light across the diameter.

CERN implemented the idea in a small machine called an 'Antiproton Accumulator', known as the AA. As its name suggests, this took antiprotons and cooled them into a well-behaved beam, gathering and storing them until there were enough to be useful. This is where van der Meer's ideas bore fruit. Electronic detectors on opposite sides of the ring monitored where the antiprotons in the beam were travelling as they passed. The signal went to a computer which calculated by how much the beams were diverging and how big a kick was needed to align them better, and then sent a signal at the speed of light to electrodes on the far side of the ring. The clever insight was that it takes about 50 per cent longer for antiprotons to travel around a semi-circle than for a signal to take the short cut across the diameter, and that if the ring was large enough, this would leave enough time for the electronics to work out what needed to be done, and to send the instructions, which the receivers act on, before the antiprotons finally arrived round the bend. In a billionth of a second, a 'nanosecond', light travels one-third of a metre, (one foot).* This was Swiss precision timing, literally and metaphorically. Every two seconds, protons burst out of the PS, smashed into a target and produced antiprotons. These entered the Antiproton Accumulator one burst at a time, where they were cooled for two seconds until the next burst arrived.

The AA was like two rings in one, connected by shutters that could be opened and closed. On the ring inside the shutters, bunches of cooled antiprotons would be circulating, while outside the shutters were the latest newly arrived antiprotons that were still in the process of being cooled. Just before the next burst was due, the shutters would open and the antiprotons in the outer ring, now cooler, switched to the inner ring. The shutters were then closed, the next burst entered, and the process repeated over and over.

Once the antiprotons were in the inner stack, van der Meer's electronic messages flashed across the ring, cooling them even

---

* This is easy to remember as it is about the size of an adult's foot, so if you can estimate how many of your feet would fit into some distance, that tells you how many nanoseconds light will take for the journey.

more. It took just over a day to accumulate and cool a hundred billion antiprotons. Van der Meer's trick led to intense beams of high energy antiprotons that could be used in experiments.

Antiprotons with their greater mass were much harder to tame than positrons, but once under control, the antiprotons packed a far bigger punch. It was this that excited the physicists. Through the annihilation of antiprotons and protons, they could reproduce in experiments conditions that had existed in the first moments of the Big Bang. For the technologists, it was the tour-de-force of the cooling, the variety of specialist race-tracks and the sophisticated electronics that amazed and illustrated that antimatter can be tamed. Yes, it is possible to make and to tame antiprotons, but it is slow, requires great patience and the price-tag corresponds to millions of dollars.

The Penning Trap

The storage of antiparticles at high energies, corresponding to temperatures that are far hotter than even the centre of the sun, involves big accelerators. Is it possible to contain them in the cold, at room temperature or below? In 1984 Hans Dehmelt managed to store a single positron for three months in an evacuated cylinder that was about half the size of a human thumb. He did so by an ingenious combination of electric and magnetic fields in what he modestly named a Penning trap, after Frans Penning, the Dutch physicist whose original idea Dehmelt cleverly built on.

The idea of the Penning trap goes back to the 1930s, the days when radios powered by valves, and televisions by cathode ray tubes, were the frontiers of electronics. Electricity flows through wires as if it is a fluid. Connect the wire at one end to the negative terminal (the anode) of a powerful battery and the other to a metal plate (the cathode) inside a glass tube filled with gas. By this means electricity passes through the gas. When this happens weird illuminations appear, and when these were first seen in the latter years of the 19th century they fascinated Victorian society. Trying to understand what was going on led to J. J. Thomson's discovery of the electron, the carrier of electric current. He did so by using electric and magnetic fields to steer the beam, and from the way it responded he realized that it consisted of lightweight electrons, smaller than atoms.

If the magnetic field is powerful enough, it can steer the electrons in tight circles, trapping them in these orbits where they are unable to cross to the far end of the tube. At least, that is what happens in a perfect vacuum; if there is some residual gas present, electrons will bump into its atoms, escape from their orbits, and current will flow.

Penning had the brainwave that this effect could be used to make a vacuum gauge. Whether current flows or is cut off depends on the voltage, the strength of the magnetic field and on the amount of gas in the tube. What Dehmelt did was to change the voltage so that current *never* flowed, but instead the electrons wandered around for ever in the magnetic field. His anode was shaped like a hollow cylinder, and the lid and base of this were split off from the sides at an angle, which he made his cathode. In effect he had made a closed canister no bigger than a soft-drink can, but instead of metals the 'Penning trap' was made from electric and magnetic fields. The first thing that Dehmelt did was to trap a single electron and measure its magnetism. The circulating electron was like a miniature radio transmitter, emitting electromagnetic radiation that Dehmelt could tune in to with a radio receiver. By measuring precisely the frequencies of radio waves, he was able to measure the magnetism of the electron to an accuracy of one per ten billion. This was far better than anyone else had ever managed and was so precise that he discovered that it was larger than Dirac's equation would have implied.

This deviation was so small, about one part in a thousand, that no one before had seen it, but it was an extremely important result. Far from it showing that Dirac's equation was wrong, it actually confirmed its profound description of the physical world. The reason was that Dirac had not just created a theory of the electron, but also of how it responded to electromagnetic fields. Richard Feynman and others had shown that the elecromagnetic field could itself turn into transient electrons and positrons, one of the many bizarre properties of quantum uncertainty. The effect of these 'virtual' particles and antiparticles in the vacuum implied that the immediate surroundings of an electron are not simply empty space but in fact are seething with activity. Dehmelt's experiment was so precise that it was measuring not just the electron but also the effects of the vacuum surrounding it. We need not get philosophical here about what exactly is an electron; it depends on how carefully you look. Look carefully enough and you will see how it disturbs the vacuum, turning the void into a hive of activity containing antiparticles. Dehmelt had proved what the theorists suspected: we live in a world of matter, but the vacuum is full of both 'virtual' antimatter and 'virtual' matter, virtual in the sense that it does not materialize (maybe that should read 'anti-materialize') but whose presence can be inferred by its effects on passing particles of matter.

That was in 1973. A decade later Dehmelt snared a positron into the trap. Holding it there for three months he was able to measure its magnetism too. All that he had to do was to reverse the direction of the magnetic fields so that the positively charged positron would experience the same constraints as the negatively charged electron had previously. And when he measured the magnetism of the positron he found that its value was elevated by the same one part in a thousand as the electron's had been. Not only was Dehmelt trapping a positron, he was able to show that it was indeed a perfect electric and magnetic mirror image of the electron.*

Antiprotons in the Trap

CERN's interest in the physics of the Big Bang had led them to make beams of antiprotons with very high energies, which involved large pieces of apparatus. To capture and store antiparticles in small-scale containers you need them to be as still as possible. So teams of scientists and engineers at CERN used their experience and built a storage ring where the antiprotons were slowed down. This Low Energy Antiproton Ring became known as LEAR. One of Dehmelt's colleagues, Gerald Gabrielse, took up the challenge of extracting antiprotons from LEAR and capturing them in a Penning trap.

The problem with antiprotons relative to positrons is their extra bulk, the resulting higher energies needed to make them, and the unwanted jiggling that ensues. CERN pithily summarized the nature of the challenge: first it was necessary to make temperatures far hotter than in the heart of the sun in order

---

* Dehmelt won the Nobel Prize in 1989; the citation reading 'for developing the ion trap which has made it possible to study a single electron [or positron] with extreme precision'.

to create the antiprotons; then to store them in the Penning trap required cooling them to temperatures colder than outer space, and with a vacuum better than you would find on the moon. The conventional way of cooling to such temperatures is to use liquid helium. However, helium is made of neutrons and protons, which would annihilate antiprotons on contact: Gabrielse wanted to store antiprotons, not annihilate them! So, like Budker, he used a gas of cold electrons as the coolant instead.

The trap was about 15 centimetres (6 inches) long. When the antiprotons entered, the voltage was increased, which created an electrical barrier like shutting a trapdoor. He captured antiprotons for the first time in 1986, and within three years was able to store sixty thousand of them for four days. But the aim was for precision, the long term storage of a few antiprotons. In 1991 a hundred antiprotons were stored for several months and by 1995 the goal of trapping a single antiproton was achieved. First, about ten thousand antiprotons and a lot of electrons from the cooling gas were imprisoned. Next, the voltage was gently pulsed, which was like opening a small window through which the electrons could escape whereas bulky antiprotons remained stuck. Then the voltage was adjusted once more so as to allow some of the antiprotons to escape until just a dozen remained. Each of these twelve circled around in the magnetic fields but at different rates. A laser beam was then tuned to kick out the fastest movers, weeding them out one by one until just a single antiproton remained. At this point the voltage was raised and the trap shut entirely. A lone antiproton was dancing in a magnetic bottle.

Having at last managed to trap an isolated antiproton, Gabrielse's team could study it at will. They compared its behaviour in the electric and magnetic fields to that of a proton. These experiments show that the antiproton and proton also are perfect mirrors of one another; their electric charges are opposite and the amounts of electric charge per unit of mass are the same to an accuracy of better than ninety parts per trillion.

Another line of research with trapped antiprotons concerned the question of whether there is antigravity. We know that matter falls to earth under gravity, and the symmetry between matter and antimatter implies that antimatter would fall towards an anti-earth. But would antimatter fall or rise in earth's gravitational field? Although no one would wager their fortune against it, the possibility that antimatter would experience antigravity could not, apparently, be entirely dismissed. If this was to be tested, the best bet seems to be antiprotons, as they are nearly two thousand times heavier than positrons. I was on the CERN committee that first examined proposals made in the 1980s by a team from Los Alamos. None of us expected that the test would be achieved as the sensitivity required was far beyond anything then available, but the challenge was so profound that we were sure it would nonetheless produce fundamental advances in antimatter technology.

Today, nearly eighty years after the discovery of the positron and half a century after that of the antiproton, everything confirms the suspicion that antiparticles are nature's mirror of particles, the yin to their yang. However, as to whether antimatter falls or rises under gravity, we still don't know.*

---

* At least, not directly. Experiments show that different materials (made of matter) all fall at the same rate. If Einstein's theory of general relativity is the correct description of gravity, this result indirectly implies that antimatter will also fall to earth at the same rate as matter.

Antihydrogen and the Antimatter Factory

Only very small amounts of antimatter can be stored in magnetic bottles. The limiting factor for positrons or antiprotons alone is that like charges repel one other, which makes it impossible to put a large quantity together because the repulsive forces between them soon become too strong for the fields in the magnetic bottle to control. In effect, the bottle will leak and the antiparticles be destroyed. Try overcoming this by putting positrons and antiprotons together to form atoms of antihydrogen, and you meet another problem. Atoms are electrically neutral and electric and magnetic fields have no hold on neutral particles—they almost immediately come into contact with normal matter, such as the walls of the vessel, and annihilate.

Before 1995 it is possible that not even one atom of antimatter had ever existed in the history of the universe. When positrons or antiprotons in cosmic rays encounter one another, they are moving so fast that they continue on their separate ways rather than lingering and combining into atoms. Everything changed that year when a team at CERN made the first handful of antihydrogen atoms.

Antiprotons circulating inside the Low Energy Antiproton Ring (LEAR) made close encounters with the atomic nuclei of a heavy element. Any antiprotons that passed close enough could both create an electron-positron pair and survive themselves; in a tiny fraction of cases, the antiproton would bind with the positron to make an atom of antihydrogen.

When the announcement of the production of nine anti-atoms at CERN was made early in 1996, the news travelled around the world to be reported in newspapers, on radio, and on television. However, the fleeting existence of the anti-atoms meant that they

93

could not be used for further studies. The feat was in having made them at all, though they lived for a mere fraction of a second before being destroyed by matter in the surroundings.

LEAR's operation ended in 1996 and it was replaced by a machine dedicated to producing and then slowing down antiparticles in order to make antimatter. In this 'Antiproton Decelerator', or AD, magnets steer the antiprotons and powerful electric fields slow them to a relatively leisurely pace, about 10 per cent of the speed of light. The AD is in fact a reincarnation of the antiproton accumulator described earlier. The only major modifications were to improve the vacuum system and to add the cooling mechanism formerly used at LEAR.

Out of each bunch of antiprotons that arrive from the AD, an experiment known as ATHENA ('AnTiHydrogEN Apparatus') captures about 10,000 of them in a magnetic cage where it slows them even more, to a few millionths of the speed of light. The next stage is to mix them with about seventy-five million cold positrons, which are collected from the decays of a radioactive isotope, caught within a second trap. Finally the positron and the antiprotons are transferred to a third 'mixing' trap. It is here that 'cold' antihydrogen atoms form.

How do the ATHENA experimenters know that they have succeeded? When a positron and an antiproton bind together to form a neutral antihydrogen atom, it escapes the trapping electromagnetic fields. The anti-atom then strikes the surroundings, the positron and antiproton annihilating separately with an electron and a proton, respectively. It is by detecting this simultaneous annihilation of the antiproton and of the positron that unambiguous evidence for antihydrogen is obtained.

In 2002 the Antiproton Decelerator made headlines when ATHENA and another experiment, ATRAP, successfully managed for the first time to produce tens of thousands of antihydrogen atoms, large enough numbers that they could begin to study a gas of antimatter. The ATHENA experiment saw its first clear signals for antihydrogen in August 2002—appropriately, the one hundredth anniversary of the birth of Paul Dirac. A month later ATRAP announced the first glimpse inside the anti-atom. The hope is that eventually it will be posssible to compare how hydrogen and antihydrogen behave in electromagnetic and gravitational fields. Any difference between matter and antimatter, however small, would have profound consequences for our fundamental understanding of nature and the universe. However, you would need to produce many billions of times greater amounts than this, and store them safely too, if you wanted to extract useful energy from antimatter, and realize the dreams of some space-travel enthusiasts. The Antiproton Decelerator is the best antimatter factory on the planet; distinguishing fact-ory from fiction will be one of the themes in our final chapter.

LEP

Antimatter was regularly made, contained, and then annihilated within the world's largest scientific instrument during the last decade of the 20th century. LEP—the Large Electron Positron collider—was indeed large. Fifty metres below the surface of Switzerland and France, in a tunnel as long as the Circle Line on the London Underground, magnets steered beams of electrons and positrons to their goal.

The raw statistics give an impression of the engineering marvel that this taming of antimatter involved. The enormous ring consisted of eight curved sections, each nearly 3 kilometres long, with 500 metres long straight sections between them. Three and a half thousand separate magnets bent the beams around the curves, and another thousand were specially constructed so as to focus the beams into intense concentrations of electric charge.

Electrons were produced by stripping them from atoms and then speeding them in a small accelerator. Positrons were made by firing the electron beam at a small tungsten target, the energy of the collisions producing both positrons and further electrons. The positrons were stored in an accumulator ring, similar in spirit to van der Meer's earlier device that had worked so well for storing antiprotons. Once there were enough, they were diverted into a series of accelerators that increased their energies, like a car moving up through the gears, until they were moving fast enough to enter into the main ring of LEP. The tubes within which they travelled ran through the centre of the magnets and formed the longest ultra-high vacuum system ever built. The insides of the tubes were pumped down to a pressure lower than on the moon; having gone to so much trouble to make, store, and focus intense beams of positrons, that was the quality of vacuum needed in order to stop them being destroyed by stray atoms of air.

The positrons sped around the 27 kilometre ring beneath Swiss vineyards, crossing the international border into France 11,000 times each second, scurrying under the statue of Voltaire in the French suburb of Geneva where he spent his final years, rushing beneath fields, forests, and villages in the foothills of the Jura mountains. For electrons there was a similar story, as the magnetic fields steered electrons and positrons on the same circular paths but in opposite directions. Keep those paths slightly apart and all is well. But at four points around the circuit small pulses of electric and magnetic forces deflected the bunches slightly so that their paths crossed. Even here the bunches were so diffuse that almost all of their individual electrons and positrons missed one another and carried on circulating. However, occasionally a positron and an electron made a direct hit, leading to their mutual annihilation in a flash of energy.

That was the key moment. The ability of antimatter to destroy matter and release all its energy was being used here by science to create, momentarily in a minute region of space, a miniature representation of what the universe as a whole was like in the first moments after the Big Bang. It is the aftermath that interested the scientists: by seeing what forms of particle and antiparticle emerged from this simulated 'mini bang' they learned how energy was first converted into substance in the real Big Bang of the early universe. Highly complex pieces of electronics encircled the collision sites, capturing and recording the emergence of these primeval pieces of matter and antimatter, as LEP repeated over and over the long-ago act of the Creation.

All this was a result of the ability to make and control beams of positrons that survived for days on end. This required engineering of such precision that LEP turned out to be affected by the motion of the moon. Initially the scientists had noticed that the electrons and positrons were arriving fractionally early, and at other times fractionally late, for their assignation of mutual destruction. The mismatch was less than a nanosecond but nonetheless LEP was sensitive to it. The timing went from early to late and back again on a twenty-eight day cycle. It was then that the physicists realized the astonishing sensitivity of the huge precision machine. The monthly lunar cycle, which raises tides of several metres in the oceans, also affects the rocks in the earth's surface, though by trifling amounts. The 27 kilometre length of LEP was expanding and contracting by a few millimetres every month, so that the beams had a little further to travel at some times, and a shorter circuit two weeks later.

Hundreds of scientists from around the world collaborated in these experiments. Needing the data instantly and having to communicate the results of their analyses to one another easily were great challenges when preparations first began in the 1980s. The World Wide Web was invented at CERN in order to do this; antimatter destroys matter but indirectly created the World Wide Web. Ten years of experiments later, LEP had shown how matter had been created when the universe was but a billionth of a second old. Out of the 'mini bang' emerged particles and antiparticles, such as electron and positron or quarks and antiquarks. Many of these were known before LEP began but it enabled scientists to understand better how these different forms of particle and antiparticle relate to one another. In addition to the familiar electron and the two varieties of quark that cluster together to make protons, neutrons, and matter as we know it, there are other varieties, rare or all but absent on earth but common in the firestorm that was the early hot universe.

It transpires that nature is not satisfied with the electron as the only possible particle for the outer reaches of the atom, but has made two heavier versions. One, some 200 times heavier, is the 'muon', while the 'tau' is 4,000 times heavier. These are identical to electrons in electric charge and, as far as we can tell, all other respects except for their greater masses. And just as the electron has its antiparticle twin the positron, so do each of these have their positively charged antiparticles.

98

FERMIONS (matter)

electron	e⁻	muon	μ⁻	tau	τ⁻	'Leptons'	Charge −1
neutrino	v_e	neutrino	v_μ	neutrino	v_τ		Charge 0
up	u	charm	c	top	t	'Quarks'	Charge +($\frac{2}{3}$)
down	d	strange	s	bottom	b		Charge −($\frac{1}{3}$)

BOSONS (force carriers)

Photon ȣ	Electromagnetic	'Electro-weak' force
W⁺ W⁻ z°	Weak	
gluons	Strong force	

Figure 8. The Standard Model of particles (quarks and leptons, which are fermions) and the carriers of the forces (which are all bosons).

A similar story happens for the quarks. Protons and neutrons are made from 'up' and 'down' quarks. There are two heavier versions of the up, known as 'charm' and 'top', and there are two heavier versions of the down: 'strange' and 'bottom'. Each of these six 'flavours' of quark has an antiquark analogue. (See figure 8.)

LEP has shown a glimpse of what the universe was like when energy was congealing into matter and antimatter in perfect harmony.* However, among the many hundreds of varieties of particle and antiparticle, a handful of examples have been found

---

* LEP has now been replaced by LHC (Large Hadron Collider) which uses protons; this is at the frontiers of technology and putting antiprotons into a future version of the LHC is currently science fiction.

where matter and antimatter, though produced symmetrically, behave asymmetrically in their brief lives and at their deaths. If we can explain how this happens, we might find clues to why matter and antimatter did not mutually destruct immediately after creation long ago, and hence why there remains something rather than nothing in the universe today. Understanding the profound relationship between antimatter and matter, and looking for subtle differences will be the next strands in our story.

7

THE MIRROR UNIVERSE

Backwards in Time?

Richard Feynman, one of the greatest theoretical physicists of
the 20th century, is famous for his eponymous diagrams. These
show the passage of particles through space and time, keeping
track of how they absorb and emit radiation, such as photons,
to interact with one another. While highly visual, they are also
primarily a code for profound pieces of mathematics that enable
physicists to calculate the behaviour of these basic components of
matter. Feynman was a young member of the Manhattan Project,
developing the atomic bomb during the Second World War. It was
while that war was ravaging Europe that, in neutral Switzerland,
Ernst Stueckelberg had produced a similar idea, albeit of limited
utility compared to what Feynman would later and independently
do. One of the implications of Stueckelberg's diagrams was that

an antiparticle could be regarded as a particle that was travelling backwards in time.

Stueckelberg's idea was published in a Swiss journal in 1941, hardly optimal for international notice at that time. Eight years later Feynman came up with similar ideas with which, and this is crucial, he was able to keep the accounts of particles and atomic physics in ways that no-one before had ever done. So powerful are Feynman's diagram techniques that they are the staple diet of all students and of calculations in physics today. Stueckelberg, however, always felt that he had not received enough credit. When asked why he had not publicized his ideas in an international journal such as the widely read American *Physical Review*, he claimed that as it was war time it had been impossible to get an artist to prepare the diagrams. This is hard to believe as they consist of little more than a few straight lines connected by wavy ones. In any event, it appears to be Stueckelberg who first proposed that antiparticles be viewed as particles travelling backwards in time.

This creates mental images of antimatter being truly exotic; that in watching positrons we are sensing electrons arriving from the future. Time and tide wait for no man; surely time cannot run backwards, and reveal this through what we have called antimatter, such that antimatter worlds are somehow coursing through our present, emerging unseen from the future, complete with anti-aliens that are growing younger by the antiday. Nor does it. To see how antimatter and time reversal relate to matter we first need to understand how the basic laws of physics care about time and also how our perception of time comes about.

For bulk matter, including living things, time is an illusion involving the laws of chance as applied to large numbers of atoms. Whereas the withering of flowers, our bodies growing older, eggs breaking and not spontaneously reassembling, and a general sense of order turning to disorder each give an intuitive sense of the passage of time, the very concept is far from obvious once one looks at the fundamental laws of physics.

Motion at all scales from planets to billiard balls is ordered by Isaac Newton's basic laws, which do not differentiate between future and past. Were we to turn back time and watch the planets in retrograde motion orbit the sun into the past, it would appear no different than if the normal affairs had been viewed in a mirror. Were we to do both of these, namely look in a mirror and reverse the direction of time, what we would see would be identical to the reality. In the jargon, Newton's laws of motion are invariant when 'P' ('parity' or mirror symmetry) and 'T' (for 'time reversal') are applied to them.

There is a manifest direction to time even though the basic equations don't care which way you run the clock* The individual atoms may care naught for time's arrow, but their mutual inter-actions, which shift them around, make a collection of atoms likely to become disordered. This is because there are more options available: there is only one way the atoms make a par-ticular egg, whereas there are countless ways its smashed pieces can fall.

As a simple example, imagine the start of a snooker game where the ten red balls are collected neatly in a triangle. The cue ball hits, and disturbs them. Every game of snooker becomes almost unique at this point as there are so many ways the balls can end up after just this opening shot. It is possible, though unlikely,

* Watch the planets long enough and even here the arrow of time will be revealed. Because of tidal friction, the moon is receding from earth, and sensors left by the Apollo astronauts can already reveal this one way movement. Wait a few billion years and the solar system as we know it will die.

that the cue ball will leave them undisturbed and itself return to the very spot from which it started. In such a case you could not tell from a film whether you were watching a real event or its time reversal. Apart from this unlikely once in a blue moon event, you could distinguish reality from a film played backwards because randomly distributed balls do not tend to collect themselves together to make a neat equilateral triangle.

The disorder of just ten snooker balls is enough to display the arrow of time. For macroscopic objects there are so many atoms involved that there is no doubt whatsoever. However, for the individual elementary particles within atoms the arrow of time is lost, as it is in the snooker game when only two balls are present. At the end of the game only the black ball and the white cue ball are on the table. It is possible to play a 'stun' shot, where the white ball hits the stationary black, stops, and transfers all its momentum to the black ball. Play that in reverse and the only thing that would give the game away is that a black 'cue' ball has hit a stationary white one. Paint them both white and you could not tell whether you were watching a film played forwards or in reverse. Likewise, at the level of individual electrons, protons, or even atoms, the laws care nothing for the direction of time.

For these individual electrically charged particles, you must do one more thing: replay time, look in a mirror and also swap the sign of electric charge everywhere. What you end up with will behave precisely as what you started with. In the analogy of the black and white snooker balls, if you also swapped black and white colours, you would not be able to tell reality from the reversal. This is the symmetry between matter and antimatter, such as between electron and positron. The mechanics of electrons and their response to forces are identical to those of positrons when viewed in a mirror and played in reverse. Thus are the electric currents due to positrons circulating anticlockwise in LEP the same as those of electrons circulating clockwise when viewed in a time-reversed movie. In such a sense does a positron behave like an electron going backwards in time.

Feynman's diagrams are a picturesque shorthand for performing calculations without getting bogged down in the profound machinery of quantum mechanics where charged particles interact with electromagnetic fields. They enable us to compute the answers; they keep track of the subtle pitfalls in accounting for positive energy and negative energy states; thinking of the latter as 'backwards in time' is very useful for avoiding some of those pitfalls but nothing, so far as we know, actually travels backwards in time. Just as electrons are negatively charged, positive energy particles moving forward in time, so are positrons their positively charged analogues, also with positive energy, and also moving forwards in time.

Put one bunch of positrons and another of electrons into LEP on a Monday and film them as they circulate into the future. At the end of the week play the movie in reverse and compare with what you had witnessed in reality. The flow of charge due to the positrons looks just like the time-reversed flow of the electrons, no more and no less than do the electrons appear like time-reversed positrons.* Positrons, like all pieces of antimatter, act no differently to the particles of ordinary matter. It is their destructive capacity, making them 'anti-matter' that tantalizes our imaginations but otherwise they are just like the more familiar players in the material world.

---

* Ignoring gravity. This idealized example should really involve transporting LEP into outer space.

The profound symmetry between matter and antimatter is revealed only when you reverse all three of the properties: charge (C), parity (P), and time (T). When Stueckelberg and Feynman did their work over sixty years ago, it was thought that it was sufficient to change any one of these, such as time alone, to expose this symmetry. However, we now know that one is not enough. Reverse just one or even a pair, and subtle differences can emerge. Matter and antimatter can be absolutely distinguished from one another.*

The Strange Behaviour of Strange Particles

The concept of primal opposing cosmic forces, the unity of opposites, is represented in Chinese philosophy by yin and yang: yin symbolizes the attributes of shade, the sinister and the left hand, while yang reflects its sunny opposite, symbolizing honesty and the right hand. Matter and antimatter share some of this mysterious symmetry which, as we shall see, turns out also to have a profound asymmetry as the symbol of yin and yang reveals. At first sight there is a perfect counterpoint in its light and dark, and yet on closer inspection the symmetry is not so obvious. (See figure 9.)

In chapter 5 we met bosons, made from a quark and an antiquark. Among the hundreds of varieties is one that turned out to display an absolute distinction between matter and antimatter. This is the strange particle known as the neutral kaon, written

---

* A profound piece of maths, known as the *CPT* theorem, shows that matter and antimatter must be symmetric when all three operations are applied. There is however a piece of fine print: this is true so long as one ignores the effects of gravity. Whether matter and antimatter behave symmetrically under the influence of gravity is the subject of debate.

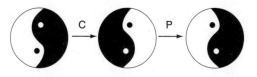

Figure 9. Yin and yang symbol compared with the mirror image of its negative.

K^0 in the particle physics shorthand. It consists of a quark and an antiquark of different flavours, whose electric charges add to nothing. A quark of the down flavour together with an antiquark of the strange variety form the K^0. Interchanging matter and antimatter, so you have a strange quark and a down antiquark, produces the anti-version of the K^0.

The first clue that the K^0 and its antiparticle were special came in 1964. An experiment at Brookhaven National Laboratory in New York was sensitive to the combination of symmetries known as CP—in effect, of swapping charges and looking in a mirror. Until that time everyone thought that matter and antimatter would behave the same: CP would be a 'symmetry' of the laws of nature. However, and to everyone's surprise, the experiment showed that this is not the case. When Jim Cronin and Val Fitch were awarded the Nobel Prize for this, one Swedish newspaper wrote the headline that the physics award that year was for the discovery that 'The laws of nature are wrong!' The laws of nature are not wrong, though they have turned out to be more subtle than anyone had expected.

Today we understand better what is going on. We have even found ways of showing the asymmetry between the K^0 and its antiparticle under timereversal.

A strange quark is more massive than a down quark but otherwise very much the same. As a result a strange quark can shed

107

some energy and convert into a down quark; likewise an anti-strange one can convert into an antidown. Conversely, if a down or antidown receives energy it can beef up to become strange or antistrange. This has profound implications for the K^0 and the anti-K^0; they keep switching identities like Dr Jekyll and Mr Hyde.

In its Jekyll form we have the combination down with anti-strange. The antistrange loses energy and turns into an antidown; sometimes this energy leaks away, causing the kaon to decay, but it can also be absorbed by the neighbouring down quark, which turns it into a strange quark. In this case what started out as down with antistrange has turned into strange with antidown: a K^0 Jekyll has changed into the anti-K^0 Hyde (see figure 10).

This means that the K^0 and anti-K^0 can swap back and forth; Jekyll to Hyde and back again. The jargon of science refers to

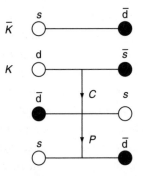

Figure 10. The Jekyll and Hyde K^0. A kaon and antikaon made of d (down) or s (strange) quarks and antiquarks (the antiquark is denoted by a line above the respective symbol). The effect of C (quark turned into its 'negative image' antiquark) and P (mirror reversal) is shown. A kaon has in effect been turned into an antikaon.

this as oscillation. If matter and antimatter are symmetric, then the change from K^0 to anti-K^0 is as likely to happen as the reverse process. A movie of the oscillations would appear the same whether it was played forwards or backwards. If matter and antimatter differ, these exchange rates could differ.

How do you look inside one of these beasts to see if it oscillates one way more than the other? The answer is to see what remains when one dies as this way you can tell if the departed was a K^0 or its anti-version. If you create a beam that is a 50:50 mixture of K^0 and its anti-twin, you can compare the mix at their deaths. If you find anything other than 50:50, it's because either there is an asymmetry in the oscillation, the transformation between Jekyll and Hyde choosing one direction more than the other, or one form dies more readily than the other. Whichever it might be, the conclusion would be that matter and antimatter differ.

A series of experiments at CERN in 1998 discovered that anti-K turned into K slightly faster than the reverse process. This proves that there is an intrinsic direction to time's arrow even at the level of the basic particles as you could tell which way a movie of the anti-K to K oscillation was playing: forwards if the anti-K tends to die out and in reverse if the anti-K emerges over time. The implication is that if you start with an equal mixture of K and anti-K, eventually a small excess of K develops. In the short life of the K this is trifling, and of itself insufficient to explain the huge dominance of matter in the universe. Nonetheless it is proof in principle that such an asymmetry can arise.

These clues to the asymmetry between matter and antimatter have come as a result of a gift of nature: in addition to the down quark and its heavier strange version, we now know that nature did not stop there. As we saw on page 99, the experiments at LEP have shown that in the early universe there were three 'generations' of quarks and three generations of electron-like particles and their associated neutrinos.

When the ideas of the Dirac equation are extended to take account of the fact that nature uses not one but three generations, it turns out that matter and antimatter need no longer be symmetric counterparts. The empirical proof of this has been seen in the asymmetric behaviour of the K and anti-K. In recent years the evidence has become even stronger with the discovery that if the strange quark or antiquark in the K and anti-K is replaced with a quark or antiquark of the third generation, bottom, the resulting bottom mesons, known as B and anti-B, show an even larger asymmetry. This is in line with what the theory predicted should occur, and proves that the existence of three generations of quark flavours in the universe enables asymmetry between matter and antimatter to arise. It also provides us with an absolute test of whether a remote galaxy is made of matter or antimatter. All that is needed is an intelligent alien to question.

Don't Shake Hands with an Anti-alien

You are hovering above some planet in a galaxy far far away, uncertain whether it is made of matter or antimatter and hence whether or not it will be safe to land. The planet is inhabited by friendly aliens with whom you have made radio contact. They are very intelligent and understand you, and being advanced know all about matter and antimatter.

Naturally they insist that they are made of matter; after all, it would be surprising if anyone chose to define their own stuff as 'anti'. How can we decide if their dictionary and ours coincide? What questions will unambiguously reveal whether they are made of the same stuff as us, or are anti-aliens?

If matter and antimatter were always perfectly symmetrically counterpoised, there would be no way to settle this issue, other than gambling with a close approach or firing a tiny unmanned probe and seeing what happens when it hits the atmosphere or anti-atmosphere.* However, we know that there is an asymmetry, small but measurable, and that is what the electrically neutral variety of K mesons can reveal. They do so when they decay, producing a pion that is either positively or negatively charged accompanied by an electron or a positron respectively. If matter and antimatter were perfect opposites, these two decays would also be precisely matched, the chance of each being the same. In reality they are slightly different.

The neutral K and anti-K are welded together in nature in such a way that they sometimes die quickly, but at other times live longer. The two possibilities are quite distinct and are known as the short- and long-lived versions. Each of these shows an asymmetry between matter and antimatter, but it is the long-lived one where the effect is biggest, the decay that leads to a positron being slightly more likely to happen than that giving an electron: out of every two-thousand examples, on the average 1,003 will give a positron and 997 give an electron. Now at last we have something to discuss with the alien.

First identify the K. It is no use giving its name, since the alien will certainly call it something else, but we can identify it by something we will all agree about: its mass. It weighs in at slightly more than half the mass of a proton or antiproton and there are

---

* Though the response of an anti-alien to a lump of matter annihilating in its anti-atmosphere makes this a risky enterprise however friendly they may have appeared to be.

no other particles that can be confused with it. So tell the alien that we are interested in a particle whose mass is slightly more than half that of the massive particle that exists in the 'nucleus' at the centre of the alien's simplest atom, the proton in the hydrogen atom (or antiproton in an atom of antihydrogen). That identifies the K.

In addition to the K^0, with no electric charge, there are also a K-plus and K-minus with positive or negative charge. So we must make sure that the alien and we are talking about the electrically neutral version. We must say that the property that holds the atom together is what we call 'charge' and that we are interested in the K that has no charge. The alien will be aware that this K^0 has two forms: one with a short life and one with a relatively long one. It is the latter that we will focus on.

Now we come to the critical bit. In our world of matter, when the long-lived K decays into a pion and an electron or positron, it is the positron mode that is the most likely. So we ask the alien: 'Is the lightweight particle that is produced most often in these decays the same as you find in your atoms, or is it the opposite?' If the alien answers that it is the same, it is a positron: the alien is made of antimatter and we should look but not touch. If the alien replies that it is the opposite, an electron, then we are all made of matter and it is safe to land.

WHY IS THERE ANYTHING AT ALL?

The Mystery of the Missing Antimatter

Antimatter is at the centre of one of the biggest mysteries: why isn't there more of it in the universe? The established wisdom is that the energetic fireball of the Big Bang fourteen billion years ago spawned matter and antimatter in perfect balance. This transubstantiation of radiant energy into particle and antiparticle is not a one way voyage; if these opposites subsequently come into contact, they annihilate one another, the energy that was previously trapped within them being liberated as gamma rays. In the dense cauldron of the infant universe such collisions would have been very common, and the newborn material would not have lasted long. If in the first moment matter and antimatter emerged equally from the Big Bang, an instant later they should have annihilated one another.

This gives another perspective on the question. The mystery is less about why antimatter has disappeared, and more a question of why has matter survived? Perhaps the answer is that there is some difference between them, that they are not perfect mirrors of one another.

In the arcane world of strange and bottom flavours we know that there are subtle differences, but the basic electrons, protons, and neutrons appear to be matched precisely by their antiparticle opposites. If there is any difference, it is beyond our ability to measure. Everything about them appears to be as Dirac predicted: particles of normal matter and their antiparticles are in perfect counterpoint.

Although antihydrogen atoms are the most detailed 'anti'-matter clusters yet seen, theory and experience imply that all atomic elements can exist in 'anti' form. As the periodic table lists the atomic elements which are made from electrons surrounding nuclei containing protons and neutrons, so will an anti-periodic table of anti-elements emerge from swarms of positrons surrounding antinuclei containing antiprotons and antineutrons. The rules of quantum mechanics that explain the stability of atoms of matter imply the same stability for atoms of antimatter. The signs of the electrical charges have been reversed, but the laws of attraction of opposites and repulsion of like charges remain the same.

The complex interactions that make amino-acids, DNA, and life will equally allow anti-elements to make everything in anti-DNA, even anti-life. The chemistry of antimatter is the same as matter: antiplanets and antimatter in all its forms are as realizable as the more familiar matter which dominates the known universe. Are antigalaxies of antistars surrounded by antiplanets of anti-matter awaiting unsuspecting astronauts in the far reaches of the universe? How sure are we that there is no antimatter at large, out there somewhere?

Earth is not like the universe at large.* We are atypical as regards the abundance of the elements, and the same could be true as regards antimatter. So it is one thing to admit that there is no antimatter hereabouts, quite another to suppose that this is true everywhere and that the entire material universe is made of matter to the exclusion of antimatter. How can we know the make-up of a distant star, seen only as a faint candle across the vastness of space? All we see from earth is the starlight and as we have no reason to suppose that the spectra of the anti-elements are any different from those of the elements, we cannot tell stars from antistars simply by looking out into the night sky.

Astronauts have landed on the moon, as have robotic probes on Mars without being annihilated, so we know that there is no antimatter up there. The whole solar system is bathed in the solar wind, the stream of subatomic particles emitted by the sun. If the sun were an antistar and the solar wind consisted of antiparticles, we would detect the gamma rays produced when these antiparticles annihilated with the matter of the planets. But we see no such gamma rays.

This also shows the nonsense of cults who believe that comets are made of antimatter.[1] As anticomets pass through the solar wind, the amounts of gamma ray emission would be enormous, each gram of annihilation liberating twice the energy of an Hiroshima-sized atomic bomb.[2] The Giotto probe successfully

---

* For example, hydrogen is rare on earth but is the commonest element in the universe. Stars like our sun are mostly hydrogen, slowly cooking the seeds of the heavier elements, but if you selected at random a volume of the universe that is millions of light years in diameter, atomic elements such as carbon, nitrogen, and oxygen, iron, silver, and gold would be all but absent.

transmitted images from inside Comet Halley. If there are anti-comets and antimeteorites, they make up less than one part in a billion of the matter in the solar system.[3]

When stars explode, their bits and pieces are ejected into space, and if trapped by the magnetic arms of our planet they crash into the upper atmosphere as cosmic rays. As the positron was discovered in cosmic rays, and as antiprotons also have been seen there, it may be tempting to think that these antiparticles are the remnants of antistars that exploded far away. On the contrary; these positrons and antiprotons are the debris formed from the energy released when a high-energy cosmic ray made of ordinary matter hits gas in the upper atmosphere. Had an antistar exploded and permeated the cosmos with anti-elements then these also would be present, but no anti-elements or antinuclei have turned up so far in the cosmic rays in earth's atmosphere. Searches for antimatter in the rays above the atmosphere are being made by the AMS satellite, and by a balloon that floats to the edge of space above the south pole.[4] However none has been seen, not even anything as simple as antihelium, in contrast to the abundance of individual positrons and antiprotons.

Perhaps these anti-elements have been destroyed en route? While this is possible there is no evidence for it. There would be distinctive gamma ray bursts coming from the annihilation of positrons by electrons in the interstellar medium, and the anni-hilation of antiprotons also would give themselves away. Admit-tedly interstellar space is nearly a vacuum but it is by no means utterly empty, so if antimatter were to travel for several light years it would bump into something sooner or later and be revealed. In addition there are millions of galaxies distributed throughout the heavens, some of which have close encounters and are distended as the tidal forces of gravity tug on their individual stars. If any of these colliding galaxies were made of antistars there would be distinctive gamma ray bursts at the boundary, but, again, none have been seen.

All signs of antimatter appear to be due to its transient creation from collisions involving ordinary matter, such as between cosmic rays and the atmosphere. For thirty years gamma rays coming from the centre of our Milky Way galaxy have signalled that there are clouds of positrons there. In 2008 'Integral', a gamma-ray telescope on a satellite, discovered that these positrons are in the neighbourhood of X-ray binary stars. These are ordinary stars that are being eaten alive by neutron stars or black holes. The gaseous material of the dying star spirals in towards the cannibal, becomes exceedingly hot and pairs of electrons and positrons form.[5] Closer to home, a large solar flare in 2002 produced high energy particles, which collided with slower particles in the solar atmosphere and created positrons. It is estimated that up to half a kilogram of positrons were produced; if that energy could be recovered by their subsequent annihilation, it would be enough to power the UK for two days.[6]

All of the evidence suggests that, with the exception of transient antiparticles produced like the above, everything within several hundred million light years of us is made of matter. This is a huge volume, to be sure, but only a fraction of the visible universe. There is still a lot of unexplored space where antimatter could dominate. Could matter and antimatter have become separated into large independent domains?

The universe as we see it today is the cold remnant of its original creation in the hot Big Bang, and when things cool, their nature can change: water freezes into snowflakes, metal becomes magnetic. Analogously, separated regions of matter and antimatter could have emerged as the universe cooled. Immediately after the Big Bang, the baby universe would have been a froth of radiant energy, matter and antimatter continuously being created and destroyed. The universe aged and cooled until it was no longer hot enough to replace the annihilated matter and antimatter with new stuff. By the natural rules of chance there would have been some regions where there was a slight excess of matter and other regions with a slight excess of antimatter. As the universe cooled further, stars and elements would emerge as the basic particles glued to one another in the matter dominated regions, and antistars would appear in the antimatter domains.

Although this remains a possibility, most models of the universe disfavour it. The received wisdom is that the entire observable universe is made of matter to the exclusion of antimatter. On the average every five cubic metres of outer space contains one proton, no antiprotons and ten billion quanta of radiation. Everything that we know about the early universe, from theory, observations and the results of experiments at LEP, suggests that in the hot aftermath of the Big Bang those numbers would have been ten billion quanta of radiation, ten billion antiprotons, and ten billion and one protons. The inference is that one of the first acts after creation was a Great Annihilation such that the matter-dominated universe today is made from the surviving one out of ten billion protons. Everything out there today is the remnant of an even grander creation.

If this is so, then something must have happened even earlier than that to tip the balance in favour of protons over antiprotons at a level of one part in billions. Something must differentiate between normal matter and antimatter. To discover what this might be, and how the imbalance between matter and antimatter first came about, we need first to understand how matter as we know it today emerged from the Big Bang.

118

Replaying the Big Bang

Not only is matter on earth not typical of the universe at large, but matter in the universe has also evolved through the aeons. On earth matter is made from atoms: electrons trapped in the cold by the electric force of the atomic nucleus. As temperature rises, atoms bump into one another ever more violently and their electrons become dislodged. Above about ten thousand degrees atoms can no longer stay in one piece. Their electrons are liberated and flow freely in a gas of electrically charged particles known as 'plasma'. This is how it is in the centre of the sun where the temperature exceeds a million degrees and its hydrogen is utterly disrupted into independent gases of electrons and protons. We can do experiments with beams of electrons and protons and see how they behave when bumping into one another at the energies typical of such a temperature. These confirm that the sun is indeed a huge nuclear fusion machine, working its way through the first stage of chemical cookery.

Experiments show that at even higher temperatures matter takes on yet more novel forms. As far as we can tell electrons stay the same at all temperatures, but protons and neutrons do not. In the cold conditions on earth, and even in the hot centre of the sun, protons and neutrons are clumps of quarks, held together by gluons. At much higher temperatures, at the limit of what can be studied in the most powerful accelerators currently available, nuclear matter seems to melt away: as atoms turn into an electrical plasma above ten thousand degrees, so do protons and neutrons turn into 'quark gluon plasma' at temperatures above about a million billion degrees.

Nowhere in the universe is that hot today, except transiently in the collisions between particles in high energy accelerators.

119

Even fifty years ago the BeVatron was able to create conditions far hotter than the sun; today we can simulate the conditions that prevailed in the immediate aftermath of the Big Bang. This is where antimatter, in the form of antiprotons and positrons, has proved the perfect tool. When protons smash into targets of matter, such as other protons, much of their energy is wasted in scattering off the material and only a fraction is available for creating new particles. However, speed antiparticles to near the speed of light, and then have them collide head on with a similar high speed beam of their material nemesis and there is total annihilation: all of the energy previously locked within their individual $E = mc^2$ is set free.

The experiments at LEP, which we met in chapter 6, confirmed that the Big Bang spawned electrons and positrons, quarks, and antiquarks, and lots of photons and gluons. That is how it was in the long ago dawn when the temperature was billions of degrees hotter than even the sun is today. As the universe aged and cooled, these basic pieces clumped together building ever more complicated structures. Trios of quarks became glued together making the permanent structures that we call protons and neutrons, and the balls of plasma that these formed, the stars, began baking the seeds of the elements. As the temperature fell further, towards what we today call room temperature, these nuclear seeds were able to hold on to passing electrons and form atoms, chemistry, biology, and life.

We have a good understanding of how matter as we know it has formed and evolved during the fourteen billion years since the Big Bang. It is ironic that we have learned much of this by using antiprotons and positrons as tools to take us back in time and see how matter was made. Had there been antiprotons and positrons in abundance in the cosmos, they could as easily have gravitated into antistars, whose cosmic kitchens would have cooked these ingredients to form anti-elements. The message is that matter and antimatter formed in matching pairs; yet only matter has managed to survive. Somewhere in the first moments of the universe, earlier than the billionth of a second that was studied by experiments at LEP, an imbalance between matter and antimatter must have emerged.

Neutrinos

In chapter 7 we learned that asymmetry between matter and antimatter is natural in a universe with three generations. When first observed for strange and then for bottom particles, this was sensational, but as more data accumulated it became clear that these phenomena involving the quarks and antiquarks are unable to explain the quantitative total dominance of matter in the universe today. Recently attention has turned to the leptons: electron-like particles and their neutral siblings, the neutrinos. What is good for three generations of quarks is also good for the three generations of leptons, and here too an asymmetry between matter and antimatter can arise, at least in theory. This is where the search for the cause of the missing antimatter is currently focused: the prime suspect appears to be the neutrinos.

Neutrinos are among the most pervasive particles in the universe, yet they are also among the most elusive. They are as near to nothing as anything that we know. Having no electrical charge, very little mass, and being able to travel through the earth like a bullet through a bank of fog, they are so ghostly that half a century after their discovery we still know less about them than other particles. Yet in recent years the suspicion has begun to grow that they may hold the key to the mystery of the missing antimatter in the universe.

Are neutrinos matter or antimatter? They have no electric charge, like a photon or a Z^0, but unlike those bosons which are neither matter nor antimatter (page 68) the neutrino is a fermion which means that it obeys Dirac's equation and cares about matter or antimatter. So what distinguishes a neutral neutrino from a neutral antineutrino?

Unlike a neutron and antineutron, whose internal make-up of quarks or antiquarks distinguishes them, a neutrino has no inner structure; it is a will-o'-the-wisp piece of spinning nothingness flitting through space at almost the speed of light. Spinning is almost all that it does, but that is enough to decide on the matter or antimatter enigma.* For fifty years this has appeared to distinguish neutrinos, matter, from their opposites, antineutrinos. But in the last couple of years the tantalizing idea has emerged that while photons (and the other bosons) are *neither* matter nor antimatter, there could be heavy versions of neutrinos that are *both*! If such bizarre entities were being formed in the cauldron of the Big Bang, their progeny would have been shared unequally among what today we call matter and antimatter.

So what's the neutrino story?

Some forms of radioactivity produce them. When a proton in a nucleus turns into a neutron, the change in energy materializes as a positron and a neutrino. Electric charge and the net number of fermions ('net' meaning the number of matter minus antimatter fermions) are conserved throughout. The positron keeps account

---

* Quantum theory implies that neutrinos can momentarily transmogrify into an electron and a W^+, and that an antineutrino can do similar, into a positron and W^-. This could give a subtle way to distinguish them but it is beyond our ability to observe in practice.

of the electric charge—there was one positive to begin with, carried by the proton, and one at the end; the net number of fermions is preserved as the antimatter positron is balanced by the matter neutrino in this accounting. Conversely, when a neutron decays, leaving a proton, there emerges an electron and an antineutrino.

If later a neutrino or antineutrino bump into matter, they give themselves away by producing the reverse process. A neutrino converts a neutron into a proton accompanied by an electron; an antineutrino converts a proton into a neutron accompanied by a positron.

Neutrinos can spin, like electrons. As we saw early on in our tale, electrons have electric charge and their spin makes them like little magnets; as they fly they can have either of two orientations, the north or the south pole pointing forwards in the direction of flight. We might visualize these as being like a corkscrew that can twist one way or the other: a right or left-handed screw. The physics jargon refers to spin just like that: we say the electron is spinning left-handed or right-handed; anticlockwise or clockwise. Neutrinos don't have electric charge, so they don't have magnetism, but the possibility of left- or right-handed spin still applies.

In the best experiments over fifty years the neutrino appeared only to spin left-handed, while an antineutrino, by contrast, spun right-handed. As the hands of a clock appear to go in reverse, anticlockwise, when seen in a mirror, so would a left-handed neutrino appear right-handed. Can looking at a neutrino in a mirror change it into an antineutrino? The first step in resolving this conundrum is to ask the following: how would we know whether it was a neutrino or an antineutrino other than by its direction of spin? Unless there is some other indicator to identify neutrino as 'particle' and antineutrino as 'antiparticle', such as watching the process that created it in association with an electron or a positron, it would be impossible to say.

At this point it is worth contemplating for a moment what we mean by 'antiparticle'. We have a sense of normal matter, containing negatively charged electrons and positive protons. The positive version of the electron is called the positron and were the negative version of the proton called a negaton, they would have appeared to be just two new particles; it is when we call them 'anti-electron' and 'antiproton' and focus on their ability to annihilate their doppelgangers that 'antimatter' begins to excite the imagination. When it comes to neutrinos we are dealing with particles that are not familiar in our material world. They pass by like apparitions, rarely revealing themselves and certainly are not trapped within matter. Instead of thinking of neutrino and anti-neutrino as pieces of matter and antimatter, and defining which is which by their affinity for electron or positron as was customary historically, we might instead say there is just a neutrino: its left-handed spin prefers electrons, while its right-handed spin prefers positrons.

For half a century it was thought that neutrinos have no mass and corkscrew through space at the speed of light. However, in the last five years we have discovered this is not true. Neutrinos emitted in conventional radioactivity, or in the fusion processes at the heart of the sun, have a tiny mass. It is so small that no-one has yet measured how much it is, but if you had some subatomic scales, it would take *at least* 10,000 neutrinos to balance just one electron.

This nugatory amount of something has huge implications. Einstein's theory of relativity implies that a fermion travelling at the speed of light remains either left-handed or right-handed; it cannot switch from one to the other. Particles that have mass, which we now know includes the neutrino, can spin either left- or right-handed, and it is possible for interactions with other particles to switch them from one to the other. So it is possible for neutrinos to spin either left- or right-handed. The same holds true for antineutrinos.

Whether these ghostly entities follow the 'left-hand matter, right-hand antimatter' rule, or whether a neutrino is 'both' matter and antimatter, in the sense that a neutrino and its antiparticle are not really distinct objects, is an open question. This possibility was pointed out by the Italian theorist Ettore Majorana, soon after the Dirac equation first appeared with its invocation of matter and antimatter. The possibility that nature uses 'Majorana Neutrinos' is one of the most exciting topics in particle physics today. One reason is that this could play the central role in explaining the origin of our matter-dominated universe.

If neutrinos had had no mass, they would still have been mysterious but nonetheless would have fitted in with the general description of particles and forces that goes by the name of the Standard Model. It is in trying to understand why neutrinos have such trifling masses, so near to zero compared to the electron and positron and yet not quite nothing, that some radical ideas have emerged. One promising line of theory is that in addition to the known lightweight neutrinos, there are very massive Majorana neutrinos awaiting discovery. These hypothetical beasts have become known as 'majorons'.

If this is true, then although majorons are beyond our reach today, they would have been created in the heat of the Big Bang along with everything else. This could have startling implications for the nature of the present universe.

If majorons have died out, the modern universe contains their progeny. According to theory the majoron, which is a massive neutral fermion, could radiate energy in the form of a 'Higgs Boson' and turn into a neutrino or an antineutrino. It could do this to any of the three flavours of neutrino, or their corresponding antineutrino, and there is no reason why it should decay to neutrinos and antineutrinos equally. This suggests a way for majorons to have beaten the apocalypse of the Great Annihilation, leaving us with something rather than nothing. Let's see how.

Apocalypse—not Quite

Immediately after the Big Bang, when the universe was very hot, the majorons would have been in thermal equilibrium—they were forming from the cauldron and decaying all the time. The universe was rapidly cooling however, and as the temperature dropped there would come a point when there was insufficient energy to make new majorons, and those dying out would no longer be replaced. The majorons died out never to reappear; only their progeny survived. It is here that an imbalanced population of neutrinos and antineutrinos would have been formed as the fossil relic of the dead majorons.

That is the critical first step, which is fine for making neutrinos, but how does this feed matter at large? The answer comes a little later in the cooling universe when quarks and antiquarks, and electrons and positrons, have been formed out of the energy. Additional quarks and antiquarks are being made by the process described earlier where neutrinos or antineutrinos collide with electrons and positrons. Very soon it will be too cold to make any more and everything will be ready for the Great Annihilation. However, let's pause for a moment and realize what the majorons have contributed. Their deaths gave birth to an imbalance between neutrinos and antineutrinos; it was in the subsequent turmoil, as the myriad particles and antiparticles were hit by the asymmetric mix of neutrinos and antineutrinos, that an excess of quarks over antiquarks emerged.

The Great Annihilation now destroys all the antimatter in a flash along with its counterbalanced pieces of matter. The progeny of the majorons have made a lopsided universe where a handful of excess quarks remain for every ten billion quarks and antiquarks that have disappeared. These survivors cool to form the universe dominated by matter, in which protons are stable (at least on the timescale of fourteen billion years) and matter as we know it exists.

This is currently the best theory of how the asymmetry between matter and antimatter came about. Experimentalists will be looking for evidence of majorons in their data in the new experiments at CERN's Large Hadron Collider, and searches in cosmic rays are also underway. However, until they are successful, this will remain an exciting but unproven theory. What is clear is that the asymmetry between matter and antimatter arose when the universe was younger and hotter than experiments can currently access. So it will not be possible to convert matter into antimatter in the laboratory by reproducing those conditions. The goal of using antimatter as a source of energy and power will require some other solution.

9

REVELATIONS

Antimatter Fictions and Factoids

Billions of years ago energy congealed into matter and antimatter. Here on earth it remained imprisoned in matter for aeons until society learned how to release a tiny fraction of it from chemicals and the nuclei of uranium atoms. It is easier to get energy out of some forms of matter than others; all we need is an efficient spark. Antimatter would be ideal, as merely to touch it releases all of the energy from anything. The problem is that antimatter is long gone from the universe, at least around here, and so before we can exploit its pyrotechnic properties we would have to make it ourselves. This is where we come up against nature's restrictions.

A fundamental truth is that creating antimatter out of energy via $E = mc^2$ always produces equal amounts of normal matter and antimatter. If you put these two amounts back together and annihilate them, you can recover that energy so long as you haven't lost any. In practice a huge amount is wasted, but even if we could make the process very efficient, we could never get back more than we put in. It is not a matter of doing more research or creating more advanced technology to find ways around these limitations: that is the nature of nature. Antimatter could only ever become a practical source of energy if we first found large amounts somewhere, analogous to oil deposits on earth.

The only hint that antimatter might still exist somewhere is if we believe the claims that the Tunguska event was the result of a lump of antimatter hitting the earth in 1908. So first, let's see what we really know about that remarkable event.

While the earth and planets are travelling around the sun on widely separated orbits that are nearly circular, they are all playing chicken on the roundabout with other lumps of rock, pieces of dead comets, and asteroids whose orbits criss-cross our own. Comets are probably the oldest members of the solar system. Consisting of gravel and ice, deep frozen ammonia and methane, many of them spend much of their time beyond Pluto in deep space where we are unaware of them. If one loops in towards the sun, the sun's warmth vaporizes the ice. The resulting ejected gases and dust reflect the sun's light so as to appear to our telescopes, and sometimes even to the naked eye, as a bright head known as a coma. The comet's centre may consist of one or two lumps of rock about a mile in diameter whereas the coma is usually bigger than the earth, extending as much as 100,000 miles across. The solar wind, high-speed particles emitted by the sun, drive fine dust particles from the coma, forming the comet's lengthy tail, which is the characteristic cometary shape as recorded in the Bayeaux tapestry and paintings throughout history.

Fragments of comets form rings of rocks, most but not all of which are found in a belt between Mars and Jupiter. Some form elongated trails around the sun and when the earth passes through one of these on its annual orbit, the rubble burns up in our atmosphere and we experience a meteor shower, such as the Leonids around 12 November and the Perseids in mid-August. Sometimes larger pieces reach ground as meteorites. The largest pieces of the comet may orbit on long after the rest have dispersed or been burned up in collisions. These extinct comet heads form some of the asteroids. The orbits of several of these cross our own, and they contain rocks that are as much as a mile in diameter.

About once in a hundred million years we chance upon a real monster. The death of the dinosaurs sixty-five million years ago is now believed to have been caused by a collision with an asteroid the size of Manhattan. Closing on earth at some 40 kilometres per second, it disrupted the atmosphere, exploded, and thundered into the ground at what today is the northern tip of the Yucatan peninsula, leaving its mark as the Chicxulub crater.

This was extreme but far from unique. Impact craters exceeding a kilometre in size pepper the planet. The famous crater in Arizona, which is over a mile wide and 3 miles in circumference, was caused by an impact 30,000 years ago between earth and a lump of rock the size of an oil-tanker. So with all of the evidence of earth being hit by rocks throughout its history and with the evidence all over the planet, what was so remarkable about the Tunguska event that led to suggestions that we had hit a lump of antimatter?

The most obvious surprise was that, had it not been for the eyewitness evidence and worldwide records of seismic activity and disturbance of the atmosphere, in other words, had this happened long ago, there would have been no permanent record to show something had actually happened at all. There was no crater; no meteoritic material from an extraterrestrial; the invader, whatever it was, had vanished in thin air. That is why Tunguska has an aura of mystery and the phenomena are all consistent with what you might expect if antimatter had hit the earth, and been utterly annihilated. As we saw on page 116, antimatter in comets is at most one part in a billion of the stuff in the solar system, so the likelihood of Tunguska being due to a lump of antirock is small. However, for those who want to appeal to the possibility of a one-off chance event, we can provide more conclusive proof. As forensic science can expose the culprit at a crime scene, so can it reveal a lot about what happened in Tunguska a century ago.

The possibility that Tunguska was indeed an antimatter strike isn't immediately dismissed as a media fancy. Scientists have carefully evaluated the pros and cons of this hypothesis, and in doing so have used the years of experience gained with antimatter at labs like CERN.[1]

These experiments showed that when antiprotons annihilate matter, the products include gamma rays and pions. That is the primary result but a secondary and important discovery is that when these hit surrounding material, lots of neutrons are ejected. So if antimatter had annihilated in the atmosphere, the surrounding forest would have been hit by a blast of neutrons, which in turn would have produced lots of radioactive carbon-14 in the trees. Studies of tree rings show how much carbon-14 is produced each year. There was nothing unusual in the Tunguska region in 1908, which runs counter to the idea that antimatter was the source of the explosion.

The favoured conclusion is that the impact was due to a piece of a comet. The dust tail of a comet, which is directed away from the sun, would have been pointed in a north-westerly direction when the comet hit at that time of the morning and explains the unusual luminescence of the night sky over Russia and Western Europe, and its absence in the United States to the east, in the week following the fall.

The radiation flash that burned the watching farmers and melted the silverware is also in accord with a comet exploding. The initial estimates showed the energy released was similar to that in a nuclear explosion, but very large chemical explosions in the air can create intense shock waves that will heat bodies when they hit them and produce the effects found at Tunguska. The chemicals within a comet, once released, react with the air and also produce energy. The net result is a vast amount of heat, a flash of light, and the set of experiences reported by the farmers. The damage was from the blast, which felled the trees, caused fires, and killed animals. The explosion of a comet would have used up its energy in the atmosphere to leave no lasting record on the surface. The lights, dust, seismic shocks are all consistent with the scenario that a comet melted explosively when it hit the atmosphere, far above the earth's surface.

There have almost certainly been many such events in the past, and with no one to have witnessed them we are unlikely to find evidence for them. The ones we do know of are the true monsters which, having done damage in the atmosphere during their fall, have enough rock left at landfall to leave a permanent imprint. For all the excitement that Tunguska created, in the course of world history it was rather small. Comets are out there and sometimes we bump into one. When we do, the results are very much like what happened in Tunguska. It has happened in the past, and will again in the future. The Tunguska event was indeed dramatic, but there is no reason to suspect that the drama implies that a lump of antimatter hit the earth one midsummer day a century ago or, indeed, ever.

The Power of Antimatter

As I watch my plants grow I don't see the carbon and oxygen atoms pulled from the air and transformed into the leaves; my breakfast cereal mysteriously turns into me, and yours into you because it is the molecules being rearranged; it is the atoms calling the tune and we lumbering macro-beings see only the large end products. As the atoms do their work, energy is liberated. The food that you ate some hours ago is turning into you, and into waste, but also producing the energy for life and keeping your body warm. Body temperature is the result of chemical reactions; it is Einstein's $E = mc^2$ at work. A small amount of the mass (the m) in your food is lost as the food is transformed and turned into energy (the E) at an exchange rate that is the square of the speed of light (c^2). In percentage terms the difference in weight between the food added to you minus any that you subsequently excrete is so small, about one part in a billion or a microgram in a kilogram, that to measure it with accuracy would require you to account for the traces of sweat and assorted DNA that every action transfers to everything we touch. It would be a hopeless task.

The conversion of one part in a billion of mass into energy is at the root of chemistry, biology, and life. It is also the source of the power of gunpowder and chemical explosives. These processes involve the electrons in the outer reaches of the atoms. However, vastly greater amounts of energy are accessible in the atomic nucleus. Atom for atom, the amount of energy released from the nucleus is up to ten million times greater than from its electrons.

So whereas chemical reactions convert just one part in a billion of matter's trapped energy, nuclear reactions can liberate up to about 1 per cent. If we could transform larger fractions of matter into energy, our ambitions would expand in parallel. In principle we could liberate the full mc^2 latent within matter into energy. That is the promise of antimatter.

The power of chemicals comes about because although each individual atom releases only trifling amounts, there are up to 10^{24} of them in each gram, each of which can contribute. Nuclear processes likewise use large amounts of uranium ore that can be dug from the ground and processed; nature locked in the energy aeons ago, and now we can liberate the 1 per cent from trillions and trillions of atoms. For antimatter there is no such possibility; as far as we know all of it was destroyed fourteen billion years ago. If you want to use antimatter you must first make every antiparticle, which is a very inefficient process. This is a fundamental restriction in nature: although the total energy is conserved in any production process, the amount of useful energy decreases due to friction and general waste. Consequently, because of these losses, only a trifle of the energy used ends up in antimatter particles, the result being that it costs much more energy to make them than can be recovered during their subsequent annihilation.

Antimatter at Large

Suppose we want grams of antimatter. Whether the plan is to bomb the Vatican as in Dan Brown's novel *Angels and Demons*,[2] or to make fuel for *Star Trek*, or for a power source, all of them beg the question not just of how to make and store it but what it should consist of. There is no need for special combustible chemistry like anti-TNT or benzene; it is the annihilation that releases energy so the simplest possible antimatter will do. We have to make it one antiparticle and anti-atom at a time. The question then is how best to assemble the stuff for storage. This is where the reality of nature rather than the fancies of science fiction begin to spoil the dreams. To make one gram of antiprotons requires nearly a trillion trillion of them; to do it with positrons requires two thousand times more. The numbers are huge. To give an idea of what this means, since the discovery of the antiproton in 1955, with LEAR at CERN and similar technology at Fermilab, the total amounts to less than a millionth of a gram. If we could collect together all that antimatter and then annihilate it with matter, we would only have enough energy to light a single electric light bulb for a few minutes. By contrast the energy expended in making it could have illuminated Times Square or Piccadilly Circus.

On the present level of efficiency, to make one gram would take a hundred thousand centuries. Of course those facilities were designed to make beams of antiparticles for specific experiments, not for storage in large quantities. Nonetheless even were we to design a machine specifically with the aim of building bulk antimatter, it is still a long way from tens of millions of years to a production line with a timescale of weeks. And when all that is done, you would still need to store it.

First, you need a high vacuum and a container comprised of electric and magnetic fields. One piece of good news is that we know how to do that and have successfully stored antiparticles in Penning traps for many weeks. However, there is a limit to how many you can keep in the bottle as problems arise when lots of charged particles are gathered in a small volume. The unavoidable fact of nature is that electrically charged particles with the same sign of charge mutually repel, so the more you have, the harder it becomes to squeeze them inside the magnetic bottle. About a million antiprotons is the largest number successfully stored, and before you get excited, that is a billion billion times smaller than what you would need for a gram. A leaky bottle of antiparticles is not the answer to my father's challenge of how to contain the ultimate destroyer.

One way to avoid this problem is to have a mixture of positrons and antiprotons, as atoms of antihydrogen. The electric charges of the positive positrons and negative antiprotons cancel one another, so the problem of too much electric charge need not worry us. We ourselves are made of billions of atoms, which contain balanced positive and negative charges such that overall we are unaware of the electrical activity within. The same could apply to antimatter. But do you see the catch? The magnetic ropes and electric walls that make the prison only confine electrically charged prisoners. If those prisoners have paired off so that their individual charges cancel to nothing, the power of the prison walls disappears like magic; to contain anything in an electromagnetic bottle needs some residual forces between the bottle and the thing. Atoms of antihydrogen are neutral to the electric and magnetic fields, and soon float off freely until, uncontained, they escape, meet matter, and are prematurely destroyed.

There are examples of bulk matter where the effects of the inner electrical charges can still be felt. The most familiar is magnetism, where the motion of electric charges gives magnetic effects; although the total negative and positive charges may annul one another, the net motion of the electrons may be like an army on the march, acting in step, so that their individual little pieces of magnetism add up. The same would happen for antimatter; as iron in our material world can be magnetic, so could anti-iron be magnetic in the anti-world. However, as the collection of lots of like-charged particles competes with the electric forces in the bottle, so would a magnetic collection interfere with the magnetic fields. It might be possible to trap some antihydrogen atoms in magnetic fields that change rapidly throughout the volume, so that the different magnetic moments of the positron and antiproton will hold the atom, but this has not been achieved even for a few anti-atoms.

Another possibility is to make atoms of positronium—a positron and an electron. This has attracted the interest of the US military in recent years.[3] Problems here are not just the fact that this atom is electrically neutral, like antihydrogen, but also that its constituents mutually destroy one another. The lifetime of a positronium atom is less than a thousandth of a second, far too short to use them as an energy storage device for a voyage to Mars. Nonetheless, the US Air Force believes in this enough to be pouring money into research. What can they be up to? It is time for revelations about the facts and fancies of antimatter.

Fantasies: Antimatter Bombs

In chapter 1 we saw the reports that the US Air Force is developing antimatter weapons. Now that we know more about the reality of antimatter, we can see that there is no possibility to make antimatter bombs for the same reason you cannot use it to store energy: we can't accumulate enough of it at high enough density. Thanks to the inefficiency of the transformation process of energy into antimatter, we do not have to worry about military applications. As an example, take a hypothetical 1 gram of antimatter.* With present technology, it would be possible to produce about a nanogram (billionth of a gram) of antimatter per year,[4] at a cost of about tens of millions of dollars. Obviously, it would take hundreds of millions of years—and in excess of US\$1,000 trillion—to make one gram. This appears ambitious even for the US military.

In addition to the cost and problems of making the stuff, storing it causes further difficulties. As we have said, 'like charges repel', so in order to contain the electric charge in a gram of pure antiprotons or of positrons, you would have to build a force field so powerful that were you to disrupt it, the explosive force as the charged particles flew apart would exceed anything that would have resulted from their annihilation. If you want to make bombs, then it seems to me that it would be better to forget the antimatter and instead exploit the power behind the technology that you would need to contain it; you don't need to go to the trouble, cost, and sheer impracticability of making antimatter as well.

What therefore are we to make of the speech by Kenneth Edwards in 2004 that excited interest in the US Air Force's research into antimatter weapons?[5]

Following Edwards' speech, newspaper reporters contacted Eglin Air Force Base, whose response was initially very positive. In July 2004 Rex Swenson at Eglin's Munitions Directorate confirmed that everyone was 'very excited about this technology'. Swenson was set to arrange media interviews with Edwards but

---

* This is the amount whose annihilation would release as much energy as a small atomic bomb; see Appendix 1.

138

within a month he was overruled by higher officials in the Air Force and Pentagon. According to the report in the *San Francisco Chronicle*, Edwards repeatedly declined to be interviewed claiming to be under strict instructions from his superiors. The tell-tale quote in the official line was that 'we're not at the point where we need to be doing any public interviews'. In the world of conspiracy theorists, this is 'proof' that the military are trying to suppress news of the latest big thing. In reality the explanation turns out to be much simpler: there were no antimatter weapons; the project was a dream.

The US Air Force was not developing antimatter weapons, Edwards' 2004 talk notwithstanding. It had however funded a small research project into antiprotons, without any secrecy, at Pennsylvania State University. I knew this project well as some years before I had been on a committee at CERN evaluating experimental projects involving antiprotons at LEAR. One of the scientists was Gerald Smith, formerly chairman of physics at Penn State and I noticed that his name was mentioned in Edwards' presentation. Dr Smith had retired from Penn State and in 2001 founded Positronics Research in Santa Fe, New Mexico. As the name implies, the emphasis has shifted from antiprotons to positrons, which are much more accessible. Advertised applications include energy storage, destruction of chemical and biological agents, nuclear medicine, and propulsion.

In 2004, when interviewed by the reporters, Smith confirmed that the Air Force had provided over three million dollars of funding for his team's research. However, there has been no demonstration or even serious claim that production or storage of bulk antimatter is achievable, not even in amounts that are the merest trifles of what would be required for energy.

One of the leading experts on antimatter, Rolf Landua of CERN, has commented on the irony that 'Scientists realized the atom bomb to be a real possibility many years before one was actually built and exploded; the public then was totally surprised and amazed. The antimatter bomb on the other hand has been imagined by the public who wants to know more about it, yet we have known for a very long time that it's not at all a practical device.'

Antimatter: To Boldy Go

The profound challenges of production and storage of antimatter have not discouraged research into the possibility of fuelling spacecraft with it. The idea is that the annihilation produces gamma rays, which shoot into a propellant heating it to huge temperatures before ejecting it out of the back of the rocket. Alternatively they can evaporate silicon carbide from a nearby surface, and the resulting gas becomes the exhaust that propels. The advantage over chemical fuels is the potential saving in weight. More than half of the weight in the Cassini-Huygens probe to Saturn was in its fuel and oxidizer tanks, and the launch vehicle weighed more than 180 times the probe itself. The propaganda for antimatter power is that to take a manned spaceship to Mars, the three tonnes of chemical propellant could be reduced to less than a hundredth of a gram of antimatter, the mass of a grain of rice.

However, the promoters of such hopes say less about the weight of the technology that will be required for containing the antimatter. Large numbers of antiprotons or positrons imply a large concentration of electric charge, which has to be contained. To store even one millionth of the amount needed for a Mars trip would require tons of electric force pushing on the walls of the fuel tank. Notwithstanding such problems, this is where NASA's hopes for spaceflight and the US Air Force's for an unmanned micro fighter aircraft have been.

In the 1990s Gerald Smith's team from Penn State University had been focused on antiprotons at LEAR. By 1997 their ambitions had moved on to producing, trapping, and transporting antimatter for rocket propulsion. In a paper[6] they outlined a possible development programme for doing so with antiprotons in sufficient amounts. This included a trap that they had designed capable of carrying up to a billion antiprotons for ten days, which was advertised as a 'prototype for a trap...capable of carrying 10^{14} antiprotons for up to 120 days, the duration of a round trip to Mars'. They remarked that they were 'confident of achieving this goal'. The strategy was to scale up to this level, though with understatement it was admitted that this would 'not be trivial', and then have a thousand such traps to transport the fuel.

This appears to have been more a management plan of how one would approach such a challenge rather than any tested proven route to a new technology. Ten years later, nothing like this has been achieved, nor was any of the work at CERN devoted to such endeavours. The maximum number of antiprotons ever stored in a trap is a million, and the focus of current research is on containing small numbers for precision measurements.

Positrons are lighter than antiprotons but easier to make. In the 1950s Eugen Sanger, the German rocket engineer, had suggested a design for a photon rocket propelled by gamma rays that had been produced by electron–positron annihilation. The idea excited science fiction but was never developed in reality, in part due to the problems of producing and storing enough positrons. However, inspired by a theory proposed by three physicists from

Germany and Massachusetts, Dr Smith is now looking into the possibility of making power sources from positrons.

To neutralize the positrons' charge without calling on antiprotons, in recent years he has been embedding them in positronium atoms at his Positronics Research Institute in Santa Fe. These normally only survive self-destruction for a microsecond but Ackermann, Shertzer and Schmelcher have predicted[7] that a particular combination of electric and magnetic fields could stretch the positronium like a dumbbell and greatly increase its chance of survival. According to Dr Smith, an implication of these ideas could be that the 'lifetime [of the stretched] positronium is [practically] infinite'.[8]

Until such time as a demonstration has been published and confirmed, it would be too early to get overexcited. The theory however is clear, and proposes that electric and magnetic fields can stretch out a positronium atom. The electric field tends to pull the electron and positron away from one another, and the magnetic field helps to hold them in place. In such circumstances their separation can be many thousands of times what is normally found in an atom and the likelihood of bumping into one another and annihilating much reduced.

So far, so good. However, it seems to me that even if this were verified someday for one or a few positronium atoms, it offers little towards developing a power source requiring trillions of electrons and positrons. To force vast numbers of them apart in this way would require electric and magnetic fields powerful enough to maintain independent clouds of positive and negative charge. With this challenge we have come full circle, and are faced with the same problem that has plagued all attempts: how to contain the large quantities of charge required for a power source? Until that is solved, stretched positronium offers nothing.

Meanwhile the US Air Force's hopes for an antimatter aircraft have been taken up by students. A research project linked to Eglin Air Force Base summarized the challenge of designing such a craft as follows:[9]

Positron energy conversion will be used for antimatter annihilation energy, which will provide the aircraft with propulsion and offensive capabilities. Eglin Air Force Base requires the following performance specifications for this prototype. The wingspan of the aircraft cannot exceed three feet. The aircraft must have a loiter capability which requires mid flight hover. To satisfy the offensive requirements paint balls will be fired from the aircraft as a safe simulated ammunition.

There in a nutshell you have the US Air Force's hopes of developing antimatter for power and weapons laid bare.

How did a speech by a funding officer, far from the centre of Pentagon power, lead the world's media to report that antimatter weapons were imminent? Edwards' speech and media briefing promoted the line that such weapons would be free from radioactive debris. Dan Brown's novel, *Angels and Demons* was at that time beginning to hit public awareness among the best-sellers, and featured an antimatter bomb that created 'no pollution or radiation'. The general assumptions that underpin Brown's work of *fiction* have uncanny parallels with what, within a year, the US Air Force spokesman and the media seemed to be promoting as fact.

Antimatter: A Fiction Thought to be Fact

Many people have never heard of CERN. Of those that have, most know it as the birthplace of the World Wide Web; fewer knew its main purpose, which is as the European Centre for experiments in particle physics. However, with the appearance of Dan Brown's novel, CERN is now famous as a laboratory in Geneva that makes antimatter. These two statements about CERN are correct; much else in the novel, which has led to much of the popular received wisdom about antimatter, is not.

Brown's book opens with a preface headlined 'FACT'. This includes 'Antimatter creates no pollution or radiation . . . is highly unstable [and] ignites when it comes in contact with absolutely anything . . . a single gram of antimatter contains the energy of a 20 kiloton nuclear bomb.' The US Air Force need read no further. CERN is credited as having created 'recently . . . the first particles of antimatter' and the curtain metaphorically rises to the question whether this 'highly volatile substance will save the world, or . . . be used to create the most deadly weapon ever made'. Having read this far, you know the answers.

These 'facts' are at best misleading and even wrong, but the popularity of Brown's novel has caused many to believe them to be true. As we have seen, antiparticles have been made for eighty years; a few atoms of antihydrogen have been made at CERN during the last decade; antimatter, in the sense of anti-atoms organized into amounts large enough to see, let alone contain, is still in the realms of fantasy and likely to remain so. Nonetheless, concerns about antimatter weapons come up in questions at almost every talk that I have given during the last five years. I doubt whether the genie can be returned to the bottle, but I hope that this chapter may help to answer those questions to the point where they no longer need to be asked.

In *Angels and Demons* the experimental production of antimatter being equated with the Creation is so central to the plot that a scientist tells the Pope the 'good news', even though it is decades old. Whatever led to our universe, it was not akin to the creation of matter at CERN, in either the fictional or the real world. It is not 'something from nothing...practically proof that Genesis is a scientific possibility'.[10] This is at best cod theology and non-science.

The Big Bang is the creation of all energy, all matter, and all of the known universe, together with its space and time. This is profound and beyond my comprehension.* We cannot recreate that singular event, but we can examine what happened afterwards, within what became our present universe.

Energy, lots of it, is what turned into matter and antimatter. Energy is not nothing; it is measurable and when you use some the power company will charge you for it. When you create antimatter together with its matter counterpart, you have to put in the same amount of energy as would be released were they to annihilate one another. You do not get matter from nothing. Now reverse the process, such that antimatter meets matter and is turned back into radiant energy. That certainly is not nothing, as *Angels and Demons* recognizes since the resulting blast is what is going to destroy the Vatican. As a demonstration of antimatter's power, the hero is invited into a laboratory in CERN where antimatter is floating in an evacuated bottle, and the scientist then demonstrates the destruction by bringing the antimatter into contact with matter.

It is at this point that some in the US military seem to have adopted this fictional work as its practical guide to antimatter, and to have ignored its many contradictions. Its preface described antimatter as the ideal source of energy which 'creates no pollution or radiation and a droplet could power New York for a day'.

---

* I tried to wrestle with this in *The Void*, Oxford, 2007.

Antimatter may not emit radiation so long as it stays away from matter, but in that case it offers nothing to bomb makers or power companies. In order to exploit this 'volatile' substance, you need to annihilate it with matter, at which it releases its trapped energy as radiation such as gamma rays. When the demonstration in the laboratory is described, the presence of this radiation is admitted because the hero is advised not to look directly at the sample and to 'shield your eyes'.[11] In reality gamma rays are so far from the visible spectrum as to be invisible, but as they can cause serious damage to cells, shielding the rest of your body would be more advisable.

The statement that there are 'No byproducts, no radiation, no pollution'[12] is ironic given that it occurs within a few paragraphs of the warning to beware of the gamma rays. The US Air Force were enthused so much that in promoting their interest in anti-matter for weapons they announced 'No Nuclear Residue'. The media trumpeted that 'a positron bomb could be a step toward one of the military's dreams from the early Cold War: a so-called "clean" superbomb',[13] not the exact wording in *Angels and Demons* but uncanny examples of fiction, written in 2000, presented as if fact in 2004.

The media reports mentioned positron weapons; what does the bomb in *Angels and Demons* consist of? The demonstration device[14] consisted of positrons; later the bomb appears to be made of antihydrogen.*

Although the interest in antihydrogen was inspired by CERN's Antiproton Decelerator, reinvented as 'an advanced antimatter production facility that promises to create antimatter in much

---

* 'Its chemical signature is that of pure hydrogen', p. 156, Brown, 2001, Corgi edition.

larger quantities', the AD actually produces fewer antiprotons than LEAR. As a major milestone in antimatter science and the state of the art in making antihydrogen, it is indeed marvellous, but trifling compared with what would be needed to make antimatter in industrial quantities. Even were it possible, the belief that antimatter technology could 'save the planet'[15] is specious. Not only do we have to use up energy to make the antimatter in the first place, but as we commented earlier, lots of energy is wasted in doing so. Created at nearly the speed of light, the antiparticles have to be tamed, which costs further energy. Many of the antiparticles are lost and all the energy used to make them is gone forever.

If we found large quantities of antimatter where nature had already expended energy in making the stuff so that we could now make use of it, our fuel problems might indeed be solved. But so long as we have to make it ourselves, we can do no better than making storage batteries, which produce less than it takes to make them in the first place. Regrettably antimatter is not a panacea for 'saving the planet'. Thankfully, neither is it 'the most deadly weapon'.

Antimatter Fact-ory

Antimatter is unlikely ever to produce large amounts of energy that would interest power companies, but in regions smaller than atoms the annihilation of antimatter has proved invaluable in medicine, technology, and fundamental science. When beams moving at near the speed of light smash head on and annihilate, the total energy is small but its concentration in a volume smaller than an atomic nucleus is huge.

All cultures have wondered about their origins, and the paradox of how something came from nothing. Why the Big Bang occurred no one yet knows, but out of its energy everything that we know was born. And it is beams of antimatter, first antiprotons and then positrons, that have enabled us to simulate the early universe in experiments, and begin to understand what it was like when less than a billionth of a second old. This is an astounding achievement of the human intellect: of groups of atoms collected together and able to think, to look out in wonder at the universe that made us, and build machines that can revisit our origins in the Big Bang. And the tool that made all this possible is antimatter. With such inspirations in fact, who needs fiction?

Appendix 1:
The Cost of Antimatter

Several times in this book I have asserted that you need this or that amount of antimatter to do something. If you are interested in the accounting, I decided to put it all here as having read this far you may be happy to do some arithmetic.

First, how does antimatter equate with the Hiroshima bomb analogy: does a gram really measure up to a 20 kiloton bomb? In fact, it is even more!

A 'kiloton of TNT' corresponds to 4 million million (4.2×10^{12}) joules (4 'TeraJoules'). A joule is a measure of energy, proportional to mass and to the square of speed; one joule is the kinetic energy of two kilograms moving at a speed of one metre per second, as non-relativistically kinetic energy $= 1/2 \, mv^2$.

A gram is one thousandth of a kilogram: 10^{-3} kg. The speed of light is 300,000 km/s or 3×10^8 metres/s. Now $E = mc^2$ so for 1 gram we get $E = 10^{-3} \times 9 \times 10^{16} \, kgm^2/s^2$ giving a total of 9×10^{13} joules or 90 'TeraJoules'. As 4.2 TeraJoules corresponds to a kiloton of TNT, then 90 Terajoules corresponds to 21.4 kiloton. That is the energy trapped within a gram of antimatter. This same amount is also trapped within a gram of matter, so we need only make *half* a gram of antimatter in order to be equally destructive as the Hiroshima bomb. This assumes however that you could liberate all of the energy at once. It is possible that after all the trouble and expense of making and storing it, the annihilation atom by atom might fizzle rather than explode.

The next aspect of antimatter is how long would it take to make a gram, or even a nanogram (ngm), a billionth of a gram?

To make a gram of antiprotons you will need 6×10^{23} of them, while a gram of positrons would require 10^{26}. The most intense source of antiprotons is at Fermilab, USA. Their record production over a month in June 2007 produced 10^{14} antiprotons. Were they able to do this every month for a year they could produce about 10^{15}, which equates to 1.5 billionths of a gram, or nanograms. Were we able to retain all of these antiprotons and annihilate them with 1.5 nanograms of matter, the total energy released would be about 270 Joules, which is like five seconds illumination by a feeble light bulb.

The CERN AD produces on average about 40,000 antiprotons each second, or about 10^{13} in a year. This is only 1 per cent of what Fermilab makes. However the purpose is different and those at CERN are colder, custom made for trapping and then capturing positrons in order to make atoms of antihydrogen. It is possible that the production rate could eventually be increased by a factor of ten or, in extreme, a hundred, but even then the world antiproton production would still only be 3 nanograms. Take all of the antiprotons ever produced in history and the lightbulb could burn for a few minutes. Even this is unrealistic as these antiprotons were lost long ago; the numbers of antiprotons that have been stored is trifling compared with this.

A physics facility at Darmstadt in Germany in the next few years might match Fermilab's production. Even after adding all of these together the total world production falls far short of the optimistic hopes expressed by those promoting antiprotons for space fuel.

As for antihydrogen, antiprotons trapped with positrons at CERN can make several hundred atoms of antihydrogen per second. To make a nanogram would take a thousand centuries. To fill a toy balloon, let alone make a gram would take longer than the universe has existed.

Appendix 2: 'The Dirac Code'

Dirac wanted to write an expression for the energy of an electron as a sum of two parts: its energy at rest mc^2 and a contribution from its motion. The latter is traditionally written pc, where p is the momentum and c the speed of light. It's not important for this description but I shall adopt it in case you want to compare with other books. Einstein had shown that these three are related by the hypotenuse rule of Pythagoras

$$E^2 = (mc^2)^2 + (pc)^2$$

What Dirac wanted to do was to write the energy E not simply as a square root of that equation, but as a simple sum consisting of some amount of (mc^2) and some of (pc) and nothing else. What he had to work out was the relative amount of each.

An easy way to see why he was attempting what appeared to be impossible is to take the familiar example of a right-angled triangle whose sides are in the ratios 3:4:5 representing mc^2, pc, and E respectively. Squaring them gives 9 and 16 which sum to 25. What Dirac in effect wanted to do was to write the energy, the '5', as a sum of the other two—some amount of '3' and some amount of '4'—and then to square this and compare with Einstein's hypotenuse relation: 25 = 9 + 16.

The unknown amounts of '4' and '3' we can call a and b respectively. So the challenge is to write

$$5 = 4a + 3b \tag{1}$$

and then to square both sides and match

$$25 = 16a^2 + 9b^2 + 12a \times b + 12b \times a \tag{2}$$

with Einstein's form

$$25 = 16 + 9. \tag{3}$$

The solution is $a^2 = 1$; $b^2 = 1$ and $a \times b + b \times a = 0$, which immediately reveals the conundrum: there are no numbers which when squared give 1 but whose product is zero! This is not a peculiarity of our choice of 3, 4, and 5; it is true whatever numbers you have for mc^2, pc, and E. In general you are trying to match the two expressions

$$E^2 = b^2(mc^2)^2 + a^2(pc)^2 + a \times b[(pc) \times (mc^2)]$$
$$+ b \times a[(mc^2) \times (pc)] \tag{4}$$

and

$$E^2 = (mc^2)^2 + (pc)^2 \tag{5}$$

and the conclusion is always that $a \times b + b \times a = 0$. The implication is that an electron's energy cannot be expressed as a simple sum of mc^2 and kinetic energy pc and also satisfy Einstein's triangle relation for E^2.

Or at least, cannot if a and b are simply numbers.

You cannot solve the conundrum with numbers but you can with matrices. If you want to know how they work, read the next section. If you are primarily interested in the implications, skip it.

How Matrices Solve Dirac's Problem

Many phenomena require more than just real numbers to describe them mathematically. One such generalization of numbers is known as 'matrices'. These involve numbers arranged in columns or rows with their own rules for addition and multiplication. Ordinary numbers correspond to having the same number all down the top left to bottom right diagonal, for example the ordinary number 1 in the language of matrices with two rows and two columns is $\left(\begin{smallmatrix} 1 & 0 \\ 0 & 1 \end{smallmatrix}\right)$ but $\left(\begin{smallmatrix} 0 & 1 \\ 1 & 0 \end{smallmatrix}\right)$ and $\left(\begin{smallmatrix} 1 & 0 \\ 0 & -1 \end{smallmatrix}\right)$ are not normal numbers.

Once we know the rules for addition and multiplication we can play with matrices as easily as with ordinary numbers. Addition has no surprises:

$$\begin{pmatrix} a & b \\ c & d \end{pmatrix} + \begin{pmatrix} A & B \\ C & D \end{pmatrix} = \begin{pmatrix} a + A & b + B \\ c + C & d + D \end{pmatrix}$$

but multiplication is less obvious: it involves the product of all elements of intersecting rows and columns

$$\begin{pmatrix} a & b \\ c & d \end{pmatrix} \times \begin{pmatrix} A & B \\ C & D \end{pmatrix} = \begin{pmatrix} aA + bC & aB + bD \\ cA + dC & cB + dD \end{pmatrix}.$$

So with this knowledge of how matrices multiply, here is an example of two matrices that would solve Dirac's problem:

$$a = \begin{pmatrix} 0 & 1 \\ 1 & 0 \end{pmatrix} \quad \text{and} \quad b = \begin{pmatrix} 1 & 0 \\ 0 & -1 \end{pmatrix}$$

You can use the above rules to verify that a^2 and b^2 each equal 1, and then to multiply $a \times b$ and $b \times a$. This is what you should find.

$$a \times b = \begin{pmatrix} 0 & -1 \\ 1 & 0 \end{pmatrix} \quad \text{and} \quad b \times a = \begin{pmatrix} 0 & 1 \\ -1 & 0 \end{pmatrix}.$$

So if a and b were these matrices, $a \times b + b \times a = 0$ and Dirac's theory works.

Negative Energy

Dirac's challenge turned out to be slightly harder, because momentum like velocity is a property of motion in any of three dimensions. You might be moving north, or east, or vertically, or in some combination of these; to describe your motion it is necessary to know the amount of speed in each of these three independent dimensions. The pc term that Dirac was dealing with is in fact three: the amount of pc in each of three dimensions. So rather than just a single quantity a there are in fact three; if we call the three dimensions x, y, z, then we need $a(x)$, $a(y)$, and $a(z)$, each of which independently must satisfy their own square equalling unity while the multiple of any distinct pair must vanish.

To understand what Dirac achieved let's first imagine that the electron had no mass, so that he only had to find $a(x)$, $a(y)$, and $a(z)$. The three matrices that solved the problem are those we just met:

$$a(x) = \begin{pmatrix} 0 & 1 \\ 1 & 0 \end{pmatrix}; \quad a(y) = \begin{pmatrix} 0 & 1 \\ -1 & 0 \end{pmatrix} \quad \text{and} \quad a(z) = \begin{pmatrix} 1 & 0 \\ 0 & -1 \end{pmatrix}^*$$

All of this would be perfect but for one thing: the electron has a mass and the above has ignored the mc^2 term. The obvious guess is that the b multiplying mc^2 is also a matrix, and this is correct, however there is a catch: with two-rows; two-columns matrices we have managed to solve the problem for the three $a(x, y, z)$ and

* Observant readers will notice that the $a(y)$ as written here when squared equals -1 rather than $+1$. Dirac's solution requires this to be multiplied by $-i$, where i is the square root of -1: i squared $= -1$ whereby the matrix squared $= +1$.

in doing so have used up all the possible independent matrices ('independent' meaning that any others are either an ordinary number, or multiples or sums of these three). In particular, our $a(z)$ is none other than what we called b before, so what can play the role of b now? The quantity b that multiplies mc^2 can only be a number, such as 1, and so we are faced with the problem that stumped us at the start: how to get rid of the unwanted $(mc^2) \times (cp)$ terms and be left with only Einstein's $(mc^2)^2 + (cp)^2$ combination? To do this, b has to be a matrix and the 'two rows; two column' matrices have exhausted the possibilities. To make it work, Dirac found he had to double everything, to matrices with 'four rows and four columns'.

This is the point where negative energy enters the story, and from it the essential step to antimatter.

A simple piece of high school algebra is all we need. As

$$\frac{1}{2}[(a + b)^2 + (a - b)^2] = a^2 + b^2 \tag{6}$$

has no term involving $a \times b$, so will

$$\frac{1}{2}[(cp + mc^2)^2 + (cp - mc^2)^2] = (cp)^2 + (mc^2)^2 \tag{7}$$

have no $(mc^2) \times (cp)$ terms and, even better, gives Einstein's $(mc^2)^2 + (cp)^2$ combination.

What Dirac did was to take his two-column matrices that had worked for the case when $mc^2 = 0$ and then applied it in duplicate, once where $b = +1$ and once where $b = -1$.

When written as two-by-two matrices these numbers are $\begin{pmatrix} 1 & 0 \\ 0 & 1 \end{pmatrix}$ and $\begin{pmatrix} -1 & 0 \\ 0 & -1 \end{pmatrix}$. Dirac then combined them in the four-by-four matrix form

$$\begin{pmatrix} 1 & 0 & 0 & 0 \\ 0 & 1 & 0 & 0 \\ 0 & 0 & -1 & 0 \\ 0 & 0 & 0 & -1 \end{pmatrix}$$

which is no longer a simple number.

In effect he had been forced to go to four-column matrices, but of an intriguingly symmetric form. The $a(x)$, $a(y)$, and $a(z)$ would sit in two opposite corners, but with the lower half having the opposite sign to the upper (this turned out to take account of the negative versus positive energy) and everywhere else would be nothing at all. In the fourth γ matrix, b becomes +1 at the top and −1 at the bottom, which is what we wrote above. It is easier to appreciate the pattern by looking at it. If you want to know how these are applied, enrol for your local physics course! At last everything worked.

Dirac's Gamma Matrices Displayed

$$\gamma(x) = \begin{pmatrix} 0 & 0 & 0 & 1 \\ 0 & 0 & 1 & 0 \\ 0 & -1 & 0 & 0 \\ -1 & 0 & 0 & 0 \end{pmatrix} \quad \gamma(y) = \begin{pmatrix} 0 & 0 & 0 & -i \\ 0 & 0 & i & 0 \\ 0 & i & 0 & 0 \\ -i & 0 & 0 & 0 \end{pmatrix}$$

where i is the square root of -1

$$\gamma(z) = \begin{pmatrix} 0 & 0 & 1 & 0 \\ 0 & 0 & 0 & -1 \\ -1 & 0 & 0 & 0 \\ 0 & 1 & 0 & 0 \end{pmatrix} \quad \gamma(4) = \begin{pmatrix} 1 & 0 & 0 & 0 \\ 0 & 1 & 0 & 0 \\ 0 & 0 & -1 & 0 \\ 0 & 0 & 0 & -1 \end{pmatrix}$$

157

Endnotes

Chapter 1

1. As reported in Cowan et al. *Nature*, 29 May 1965, p. 861.
2. *Energy Consumption in the UK*, dti report, page 8 gives the consumption in kilograms of oil; see <http://www.berr.gov.uk/files/file11250.pdf> 1kg of oil = 5.3×10^7 Joules.
3. Keay Davidson, *San Francisco Chronicle*, 4 October 2004. Edwards' presentation promoted 'No Nuclear Residue' in prominent red letters on several occasions and can be found at <http://www.niac.usra.edu/files/library/meetings/fellows/Mar-04/Kenneth-Edwards.pdf> An example of the paranoia induced by this episode can be found at <http://www.circling.org/archives/000022.html> which also links to other sites in this genre.
4. See Appendix 1.

Chapter 3

1. *The Longman Literary Companion to Science*, Longman, Harlow, 1989, p. 170, as reported in P. Coveney and R. Highfield, *The Arrow of Time*.
2. P. A. M. Dirac; Proc. Roy. Soc., September 1931. As quoted in Fraser, *Antimatter*, p. 62.

Chapter 4

1. D. Wilson op. cit., p. 548.
2. *Nature*, 2007, vol. 449, p. 153.

Chapter 8

1. For example see <http://www.matter-antimatter.com>.
2. See Appendix 1: The Cost of Antimatter.

3. D. Fargiana and M. Khloper, *Astroparticle Physics*, 2003, vol. 19, p. 441.

4. AMS (Anti Matter Spectrometer). See also 'The Hunt for Antihelium', *Britannica* online, *Science News*, 21 May 2007.

5. G. Weidenspanter et al., *Nature*, 2008, vol. 451, p. 159.

6. See chapter 9 and *Energy Consumption in the UK*, op. cit.

Chapter 9

1. The forensic science is due to Cowan, Atluri, and Libby, *Nature*, 29 May 1965, p. 861.

2. Dan Brown, *Angels and Demons*. Corgi, 2001.

3. For example, see < http://www.niac.usra.edu/files/library/meetings/fellows/Mar_04/Kenneth_Edwards.pdf>.

4. See Appendix 1: The Cost of Antimatter.

5. See chapter 1 and < http://www.niac.usra.edu/files/library/meetings/fellows/Mar_04/Kenneth_Edwards.pdf>.

6. 'Production and trapping of antimatter for space propulsion applications'; M. Holzscheiter et al.; Penn State report; <http://www.engr.psu.edu/antimatter/documents.html>.

7. J. Ackermann, J. Shertzer, and P. Schmelcher, *Physical Review Letters*, 1997, vol. 78, p. 199; and *Physical Review*, 1998, vol. A58, p. 1129.

8. As reported by Mark Anderson, *National Geographic News*, 4 May 2006.

9. See <http://www.eng.fsu.edu/ME_senior_design/2004/team7>.

10. Brown, 2001, Corgi edition, p. 105.

11. Ibid., p. 103.

12. Ibid., p. 102.

13. *San Francisco Chronicle*, 4 October 2004. Not the exact wording as in *Angels and Demons*.

14. Brown, 2001, Corgi edition, p. 98.

15. Ibid., p. 106.

Bibliography

Close, F. *Too Hot To Handle: The Race for Cold Fusion*. W. H. Allen, 1990.
—— *The Void*, Oxford, 2007. .
Coveney, P. and R. Highfield *The Arrow of Time*. W. H. Allen, 1990.
Dirac, P. A. M. *The Principles of Quantum Mechanics*. Cambridge, 1998.
Energy Consumption in the UK. <http://www.berr.gov.uk/files/file11250.
  pdf>.
Fraser, G. *Antimatter: The Ultimate Mirror*. Cambridge, 2000.
Vonnegut, Kurt *Cat's Cradle*. Holt, Rinehart & Winston, 1963.
Wilson, D. *Rutherford; Simple Genius*. Hodder and Stoughton, 1983.

Index